Y0-BEE-520

The Fascinating Universe

Werner Buedeler

The Fascinating Universe

The Modern Aspects of Astronomy

Translated by Fred Bradley

Van Nostrand Reinhold Company
New York Cincinnati Toronto London Melbourne

Library of Congress Catalog Card Number 82-10912
ISBN 0-442-21427-8

Printed in the United States of America
Designed by Loudan Enterprises

Published by Van Nostrand Reinhold Company Inc.
135 West 50th Street
New York, New York 10020

Van Nostrand Reinhold
480 Latrobe Street
Melbourne, Victoria 3000, Australia

Van Nostrand Reinhold Company Limited
Molly Millars Lane
Wokingham, Berkshire, England RG11 2PY

16 15 14 13 12 11 10 9 8 7 6 5 4 3 2 1

Library of Congress Cataloging in Publication Data
Buedeler, Werner, 1928-
The fascinating universe.
Translation of: Faszinierendes Weltall.
Bibliography: p.
Includes index.
1. Astronomy. I. Title.
QB43.2.B813 1983 520 82-10912
ISBN 0-442-21427-8

Contents

Our Sun—154

Other Suns—198

The Universe in Space and Time—230

The Universe in History

THE POSITION OF MAN IN THE UNIVERSE

During the last few decades a change, unnoticed by most of us, has taken place in the science of the stars. In the past astronomy was simply observing through a telescope, watching sunspots move across the sun's disk, predicting solar and lunar eclipses, and taking note of a star whose brightness changed regularly. Also covered by the discipline of astronomy were measuring the motion of double stars, or binaries, analyzing the sun's light with a prism—which disperses it into a band of colors, (the spectrum), producing drawings of the appearance of the planets Jupiter and Mars—photographing spiral nebulae; counting stars; and establishment of the time from the apparent revolution of the sky, the result of the Earth's rotation around its axis. It is true that all these activities still form part of astronomy today, but—even when they are still performed at all—they have become rather marginal pursuits.

Astronomy as we now know it today consists mainly of physical and mathematical analysis and evaluation of data automatically transmitted by and recorded from balloons with astronomical instruments launched into the upper atmosphere, artificial earth satellites, and space probes sent to explore other planets. It involves the recording of cosmic radio signals by means of dish-shaped aerials that are directed at the sky like ears, listening for signals that will tell us something about the structure of the universe in regions we cannot see because they are obscured by dark matter. To the telescope, the "long-distance eye," the optical instrument that enables us to see and to photograph the stars, the "long-distance ear" has been added, an instrument that records noise. Pulses within certain wavelengths beyond the radio spectrum are the cosmic music for the ballet of the molecules and atoms, for the fusion and fission of particles once but no longer considered the elementary components of matter.

The relative importance of astronomy has also changed. It is no longer an unprofitable occupation. Today astronomers, like physicists, chemists, industrial researchers, and structural engineers of giant projects, budget in terms of millions of dollars. As nuclear physicists are building enormously powerful devices for the acceleration of protons, electrons, and a veritable zoo of short-lived subatomic particles, astronomers are building huge telescopes, balloon gondolas with automatic recording devices, and satellites and vehicles that move about on their own on the moon and on Mars and are capable of automatically transmitting to earth information gathered on their missions.

Astronomy is a discipline that calls for more flexibility, more adaptability, more interdisciplinary knowledge and ability than many other fields of scientific research. To be an astronomer is to be a physicist and a mathematician, a chemist and an electronics expert, and often even a biologist and a geologist. The modern astronomer has stopped spending most of his time at the telescope at night —many scarcely use it at all. Rather, they sit at their electronic computers from morning until night, developing theories or hunting for numerical values for radio signals, displacements of spectrum lines, or densities of star clusters.

The questions the astronomer is now expected to answer are also different. Some of the questions were formulated thousands of years ago by Greek philosophers; however, others can be understood only within the context of our own age. But questions that were the subject of controversy between astronomy (or philosophy) and the Church during the Middle Ages and the Renaissance have lost their substance as a result of the modern approach, because not only has the Church changed its opinions but astronomy, too, is moving closer to religion—questions and answers often tally in both fields.

In 1978 a book was published entitled *God and the Astronomers,* by the well-known astronomer and space scientist Dr. Robert Jastrow. The starting point of his reflections is the theory, borne out by observation, that our universe had a beginning in space and time, that it began with a primeval explosion, presumably 15 to 20 billion years ago, and that it is basically pointless to ask what had existed before. No matter how we turn and twist, the concept of the Creator, the concept of God, is necessarily introduced, because our inability to determine what happened at the moment of the Big Bang is fundamental. This, Jastrow writes, is "the crux of the new story of creation."

During the last half century many astronomers and physicists were irritated by the thought of a universe that originated at a fixed point in time. This theory was first formulated because of a discovery by the American astronomer Vesto Melvin Slipher in 1913, the fact that some galactic systems recede from the earth at velocities of up to almost 1,000 km/sec (3,600,000 kmph). In 1917 the Dutch astronomer Willem de Sitter postulated—based on Einstein's General Theory of Relativity—an expanding universe in which all galactic systems are moving away from each other. He was obviously unaware of Slipher's observations. The idea of an expanding universe that came into being at a fixed point in time finally gained a foothold around 1930 after Edwin Hubble and Milton Humason had shown the recession of the spiral nebulae to be real, the Russian Alexander Friedmann and the Belgian Georges Lemaitre had underpinned it theoretically, and the famous British astronomer Sir Arthur Eddington had also pronounced himself in favor of an expanding universe. However, the decisive evidence was not provided until 1965 when Americans Arno Penzias and Robert Wilson discovered a cosmic radiation that arrived from all directions in space; this had been predicted in 1948 by Ralph Alpher and Robert Herman as the residual radiation from the origin of the elements in a small, unimaginably hot universe. On the question of the "before," Jastrow remarks drily in his book that most physicists and astronomers today tend to adopt St. Augustine's answer to the question of what God did before He created heaven and earth: "He created hell for those who ask such questions!"

The question of the act of creation, however, is not the only one in which we can discern a convergence of astronomy and religion; many other old-fashioned concepts of medieval astronomy are returning in a new guise. Today we know beyond a shadow of a doubt that the sun is the center of our planetary system and that the Ptolemaic geocentric system (which assumes that the earth is the center of the universe) is wrong. But at the same time we are increasingly

This Egyptian altar relief (circa 1335 BC), like many similar pieces, indicates the role the sun played as a god. His rays blessed mankind, helped people to obtain gifts from the gods, and determined the course of events. Out of the sun cult developed the determination of the sun's orbit and Egyptians gained a far-reaching knowledge of its motions.

forced to accept the anthropocentric approach of regarding the earth as a center of *life* (at least in our own planetary system), because probes in our own planetary system and throughout the universe have failed to pick up signals from other intelligent beings. The hopes placed in the Viking space probes that landed on Mars have been dashed. After all the seemingly contradictory data received from the two probes about the surface of Mars have been analyzed, we must concede that the likelihood that we shall find any traces even of primitive life is almost nil.

The astronauts who landed on the moon between 1968 and 1972, too, have experienced the loneliness of space and come to appreciate the value of our planet. Frank Borman's observation on

Christmas 1968, on his way to the moon, 200,000 km out in space, on "Spaceship Earth" expresses his deeply felt love and affection for his native planet. The moral of his parable is a concern about the conservation of our environment. He stated that, just as food, oxygen, energy, and filter material for air purification were limited on his spacecraft so that he and his fellow astronauts had to husband their resources, so the resources of the earth, also a vehicle traveling through space, are finite and must be used sparingly if mankind is not to incur the danger of its own demise.

Borman's comparison has often been quoted, and his photograph of the earth from a distance halfway to the moon published a thousand times—and both picture and parable are certain to have contributed to the heightening of man's awareness of the environment. Naturally, some conclusions have been drawn that may not be entirely correct from an objective point of view, such as those concerning the dangers of nuclear energy. Nevertheless, the photographs of the earth from space can be analyzed through special techniques to reveal evidence of minerals and water supplies and of environmental pollution. Volcanoes and minute details of mountain ranges can be identified; even the vapor trails of airplanes can be made out. Only one of the most marked manifestations of human life is invisible: the frontiers that separate nations, though these can sometimes be distinguished indirectly. A striking example is an infrared photograph taken during the Apollo 9 mission of the Salton Sea area of the United States and Mexico. The frontier between the two countries stands out as if drawn with a ruler: in the United States we see well-irrigated, luscious fields, whereas no irrigation canals have been dug in Mexico; as a result the land is as arid as a desert.

Unfortunately, the history of mankind has taught us that intellectual perceptions require centuries to imprint themselves upon the collective consciousness and to become universally accepted. Good examples are the fact that the earth is round and that it revolves around the sun; presumably and regrettably the necessity for environmental conservation and for the elimination of political boundaries will suffer the same fate. What has been done so far to solve the energy crisis is limited to insubstantial and ineffectual political proclamations. No steps have been taken to follow the thoughts of Copernicus in a Copernican world: every nation (or at least its government) is still obsessed by the political illusion that it is the navel of the world. The conflicts that have constantly occurred since the First World War are ample evidence of this state of mind. Wars are not a disease but a symptom.

The latest findings of space research also teach us that we must give up the Ptolemaic mentality regarding the planet earth as a whole. The investigation of interplanetary space, of the sun, the moon, and other planets, shows us more and more clearly that the earth cannot be considered in isolation from the universe, that it has numerous interrelations with other heavenly bodies and above all with the sun. The earth is part of the universe and cannot be considered apart from its cosmic environment.

It is certain that the relations between the sun and the earth (that is, the effects of solar events on the earth) are much more far-reaching than we are aware of today. The moon, too, exerts a number of influences on the earth and on man that have yet to be investigated in detail. Only within the last few years have these phenomena become subjects of research, thanks to the readiness of the various scientific disciplines to cooperate more closely. This statement is not meant to advocate nonsensical astrological superstitions, unfortunately still rampant (even among framers of public opinion), but instead to draw attention to the interactions between physics, physiology, and psychology, the study of which is still in its infancy.

ASTRONOMIC AWAKENING

Our current idea of the world is the result of thousands of years of wrestling with reality. The first ideas man had about the universe are hidden in mythical pictures of the world in prehistoric times. Some of them are reflected even now in ancient superstitions or misinterpreted observations of nature. The early concepts about the structure of our world began with the earth itself, as yet unexplored. Hypothesis or fantasy therefore had to make up for the lack of concrete knowledge and so the frequent embroidery of fairy-tales, stories, and superstition.

Man's interest in the sun, the moon, and the stars certainly goes back to the time when he became settled. The sun was important to him not only because it divided the fearsome hours of darkness from the warm and radiant hours of the day. But the sun gained significance for man as soon as he took up agriculture. For the annual path of the sun dictates the sequence of spring, summer, fall, and winter.

Man learned to grasp the connections between the seasons: at a certain time of the year he observed that the weather began to turn cold, and he remembered that he had experienced such a period before. Soon he anticipated the seasons, and during the fall, when the trees shed their leaves, he gathered the firewood needed during the cold winter months. Man noticed that in summer days were long, the sun rose far in the east and set far in the west, and during the winter the days were short, as was the diurnal arc of the sun. So at a very early stage man must have suspected a deity in the warming, life-sustaining sun. We know of many ancient cultures that had a sun-worshiping religion. It developed in Babylon, in Egypt, in the South America of the Incas, and in other regions. Sun temples were erected, and the sun god was worshiped.

The science of astronomy began with the observation of the sun and the moon, but it soon progressed to the fanciful combination of stars into constellations. The constellations, as did the sun and the moon, served as guides through their annual progress across the heavens. It thus became possible to measure longer periods of time and to determine the beginning of a new season. As a result the sun's annual motion was made the basis of the calendar as its diurnal motion (the rotation of the earth on its axis as we say today) became the basis of the day.

The moon, the celestial body from which the division of the year into months was derived, confused rather than clarified the situation. The difficulty began with the fact that the year is not an integral multiple of the moon's orbits around the earth. The moon was observed to take twenty nine to thirty days from one full moon to the next, that is, for one orbit. But a year consists of about 365 (exactly 365.2425) days, or about 12.38 lunar orbits, not of an integral number of months. This is the crux of the problem: in a pure lunar year the seasons are displaced.

The lunar year was introduced by the Greeks and the Romans and initially consisted of twelve lunar months of 29.5 days each, totaling 354 days. Mohammed, too, introduced a lunar year and even now it is used by the Jews and the Mohammedans in the form of the Lunar Cycle to fix their religious holidays. The Lunar Cycle is an arrangement whereby the lunar year is made to coincide with the solar year through intercalation. It is a well-known fact that in our own calendar the moon still determines certain dates; for example, Easter Sunday is always the first Sunday after the first full moon following the spring equinox.

It is obvious that even during that early period some people would not be content simply to accept the sun and the moon as deities and to explain the stars as heavenly ornaments. But this acceptance was the prevalent attitude. The ancients' motives for observing the sky were religious veneration of the stars (and its adjunct, astrology) and the establishment of a calendar. The reason for the motion of the stars was of concern neither to the Egyptians and Babylonians nor to the Indians, Aztecs, Mayas, Chinese, and Persians.

We do not know for certain when and where the first astronomic facts were discovered and observed. These include the realization that the sun rises in the east and sets in the west, that the diurnal arc

Through this refracting telescope at the Wilhelm Foerster Observatory in Berlin many laymen can observe events on the sun. One aim of this observatory is to spread an understanding of astronomy among the general public.

is wider and higher during the summer than during the winter, that the moon passes through phases from the new moon to the full moon. We do not know how close the contacts were between the nations about 3000 or 4000 BC, but this is of little importance because these items of elementary knowledge were surely discovered in several places independently. The motion of the planets among the seemingly fixed stars, in relation to each other, cannot have escaped the careful observer.

The fanciful combination of individual fixed stars into constellations must be associated with that elementary complex of knowledge which, although it indicates interest in the celestial objects, does not yet amount to a science of astronomy. The stage for this was not yet set.

Early Calendars

The Egyptians used several concepts of the year. The most popular one consisted of twelve groups of thirty days, with five additional days added at the end of the year. But because the year has 365.25 days, under this system each new year began one-quarter day earlier than its predecessor, adding up to one day every four years, ten days every forty years, one hundred days every four hundred years, and so on. This meant that, over the course of the centuries, the beginning of the Egyptian year advanced through the solar year and with it the seasons.

A good indication of time measurement with the aid of the stars comes from First Dynasty Egypt (2800 BC). Sirius, the main star of the constellation Canis major, the Great Dog, and the brightest star in the sky, was called Sothis by the Egyptians and worshiped as the "Bringer of the New Year and the Flood." It was also the basis of another calendar, the Sothis year. It must already have existed at the time the moving year with its 365 days was introduced. The So this year was a fixed year and divided into three seasons: the "months of inundation," the "months of growth," and the "months of heat." Each of these seasons had four months; the New Year was fixed on a day shortly before the annual Nile flood.

The Nile bursting its banks was a special event to the Egyptians, who were farmers. The Nile flood was heralded by a conspicuous astronomical event, the first sighting of Sothis in the morning sky after a season of Sothis not appearing. (This event is called heliacal rise. The term is derived from *helios*, the Greek word for "sun.") At the time in question the heliacal rising of Sirius took place about July 20, every year. Because of the precession of the equinoxes (the gyration of the earth's axis) over 26,000 years, Sirius now reappears during the first days of July.

Heliacal Rising and Setting of Stars

If we look at the constellations directly above the point of sunset shortly after the sun has disappeared below the horizon, we note that they set progressively earlier and eventually are no longer visible. Gradually, they are replaced by other constellations moving in from the east. The original constellations disappear in the rays of the sun; likewise, new constellations emerge in the course of time from the rays of the sun in the morning sky shortly before sunrise.

In fact, all except the circumpolar stars pass through this annual cycle of heliacal setting and heliacal rising. Circumpolar stars are those located near the celestial pole and therefore describe their entire annual path above the horizon. At a mean U.S. latitude of 42° and U.K. latitude of 54° their angular distance from the celestial pole (marked approximately by the Pole Star) is less than 42° and 54° respectively. All other stars are invisible at certain times of the year. As we can easily observe during a long enough period of time, a star rises four minutes earlier from one night to the next, and accordingly sets four minutes earlier.

Today, we know that this phenomenon is caused by the earth's motion round the sun. Those constellations which happen to be invisible are—seen from the earth—in line with the sun, therefore rising and setting with it; they cannot be seen because they are swamped by the sun's brightness and the scatter of its light in the earth's atmosphere.

The constellations that happen to be in line with the sun can, of course, be seen from the moon (which has no atmosphere) or in empty space, because here the stars are not swamped by the light from the sun and can therefore be seen together with it.

The Egyptians did not confine their observation of the heliacal rising to Sirius by any means, although Sirius occupied a special position as the star of the beginning of the New Year and of the Nile flood. In addition to the solar year the Egyptians had thirty-six ten-day weeks into which they divided their year. Each of these weeks was associated with a star or a group of stars and began with the heliacal rising of these stars or star groups. The weeks were called decans and went beyond the end of the year without the addition of intercalary days. Their significance was above all mythological and astrological. Deities said to have influence upon the fate of the individual were associated with the decans.

Numerous astronomic representations and inscriptions have been found in the temples and burial chambers of Egypt, indicating great antiquity of astronomic knowledge. The frequent conclusion that the Egyptians had a very highly developed astronomy is incorrect. Egypt's astronomy, pursued by the priests, was largely confined to the religious worship of the stars and their observation for calendrical purposes. The Egyptians had no knowledge beyond this, especially in the sense of a cosmology, a science of the structure of the universe or of the solar system. Three thousand to five thousand years ago they imagined the earth to be a flat quadrangle traversed by the Nile and not extending beyond the part of the Orient then known. Massive pillars marked the four corners, supporting the vault of the sky, from which the stars were suspended like lamps.

Compared with the astronomy of the early Egyptians, that of the Babylonians was very much more advanced, although even the Babylonians did not yet have a cosmology that one could call realistic. But the priests of Babylon observed the motions of the sun, the moon, and the planets far more meticulously than the Egyptians did, and against a very pronounced astrological background. Thus many records of the motions of the stars have come down to us in the form of so-called *omina*. These are events observed in the sky that indicate their presumed importance to the good or bad fortune of the country. The ancient Babylonian astrologers did not connect personal fate to the motions of the stars; according to their interpretation the stars merely influenced events on a large canvas, war or peace, good or bad harvests, floods, droughts, coronations of kings and the like.

By 700 BC the Babylonians were able to observe the risings and settings of the stars, times of culmination (the highest point reached in the sky), heliacal risings and settings, motions of the planets, and phases of the moon with reasonable accuracy. They used a lunar calendar; each month began with the appearance of the new crescent moon in the early evening sky. By linking the lunar and the solar year in a nineteen-year cycle they adjusted the lunar year of 354 days, divided into twelve months of alternately twenty-nine and thirty days, to the changes of the solar year. This adjustment initially consisted in the occasional intercalation of a thirteenth month as the need arose, but later became systematic. This lunisolar year was founded on the realization that twelve "short" years of twelve months and seven "long" years of thirteen months corresponded with only very little difference to 235 lunar months of 29.5 days each. Solar and lunar eclipses, too, were already very accurately observed by the Babylonians, who discovered that these events mostly recurred every eighteen years and eleven days, the so-called Saros Period.

INSTRUMENTS FOR DEFINING TIME

The precision the Babylonians attained in their celestial observations is astonishing in view of the fact that the aids they had available for this purpose were very primitive. They made frequent use of the gnomon, a simple, vertical column or shaft. By measuring the shadow cast by the gnomon, true midday (the time at which the sun reaches its highest point in the sky and thereby passes the median line of its diurnal arc, the southern meridian could be determined). Continual observation (over weeks and months) indicated the time of the summer and winter solstice. The Babylonians used the gnomon to determine, for instance, the length of the year to an accuracy of 4.5 minutes.

There are other well-known definitions of the length of the year: the tropical year and the sidereal year. The *tropical year* is defined as the span of time between two passages of the sun through the vernal, or spring equinox, the moment when the sun crosses the celestial equator from south to north. In a leap year this occurs on March 20, in other years on March 21. The *sidereal year* is determined by the time the sun takes in the sky to return to a certain star after a complete orbit. Because of the previously mentioned precession the tropical is about twenty minutes shorter than the sidereal year. Expressed in fractions of a day the tropical year has 365.2422, the sidereal 365.2564 days. The obvious reason for the difference is that the earth's axis precedes in the opposite direction of the earth's orbit around the sun. The point of the vernal equinox therefore moves toward the sun on the apparent orbit of the sun. The sun thus need not move full circle in its annual orbit in the sky to pass from one vernal equinox to the next. Although the Babylonians determined the length of the tropical year to an accuracy of 4.5 minutes, neither they nor any other contemporary nation knew about precession and the difference between the tropical and the sidereal year.

Besides the gnomon, which the Babylonians and later the Egyptians used for these measurements, there were only sundials, water clocks (clepsydras), and various sighting devices. The gnomon, the forerunner of the sundial, is probably the oldest instrument used for the determination of the sun's altitude and, thereby, time. But the length of the shadow cast by a person's body, too, served this purpose. It was measured with the person's own foot; this compensated for peoples' different body heights, because taller people also have longer feet, so that the suggestion to meet "when the shadow measures sixteen feet" roughly meant the same time for everybody.

Later, sun pillars were erected in many places; the length and position of their shadow became the daily time measure. The pyramids, aligned with the four directions of the compass, could also have served as chronometers, although the many quantitative relations to astronomical events that were attributed to them probably belong to the realm of fantasy.

In 1500 BC, about 1200 years after the first pyramids had been built in Egypt, the Pharaoh Tuthmosis III erected a sundial on which he immortalized his signature. It was a long horizontal board with another board, the gnomon, mounted at one end. The horizontal board was graduated in hours. Tuthmosis's sundial always had to be turned in the direction of the sun, so that the sun's shadow was cast parallel to the time markings. It indicated the early hours of the morning and evening with remarkable accuracy. Less accurate were its readings around midday, when the sun's altitude changes little. No readings were possible at the time of sunrise and sunset, when the length of the shadow cast by the gnomon is infinite. Tuthmosis' was a type of sundial whose time measurement was based exclusively on the sun's altitude, disregarding the sun's azimuth, that is, the direction of the compass in which the sun happened to be.

This allows the conclusion that the Egyptians already divided the day into twelve hours. However, the hours were of different length because of the change in the length of the sun's diurnal arc over the course of a year. But in the Egyptian division of time, one hour always was one twelfth of the time between sunrise and sunset. This was accepted; in those days people adapted themselves to the variations of nature, unlike today, when nature is squeezed into an abstract framework determined by clockwork mechanisms or oscillations of crystals.

The twelve hours of the night were also subject to the same fluctuations in the rhythm of the year as the daylight hours. Although they remained of equal length in any one night, in the course of time their length changed inversely to that of the daylight hours: the hours at night became shorter as the hours during the day became longer and vice versa.

But fixing the hour at night was fraught with difficulties. To read it from the stars it was necessary to know the sky, a knowledge that was by no means universal. Starlight is too weak to cast any shadows, so a gnomon was of little use. To overcome this difficulty, two types of water clock were invented almost simultaneously.

The flowing type of clepsydra consisted of a vessel that was filled at sunset. It had an opening in the bottom, through which the water slowly escaped, so that at sunrise the clock was empty. The inside of the vessel was graduated in hours. Soon refinements were introduced: in the first of these water clocks the time marks inside the vessel were unevenly spaced, because the emptier the vessel, the slower the flow of the water owing to the reduction of water pressure. The hour marks that indicated the approaching morning were therefore more widely spaced than those before midnight. But this flaw was soon ironed out; a vessel of parabolic shape was introduced to make the hour graduations uniform. It widened from the bottom upwards so that the water level dropped evenly.

In the second type of clepsydra, water from an overflowing vessel was conducted into a tank with a small hole through which the water entered and slowly filled the water clock. On one side was the graduation for the longer summer hours, and on the other marks for the shorter winter hours. The hour marks were therefore curved. Such clocks soon became part of the equipment of upper-class households. During the third century BC the sand glass was invented; this had the obvious advantage of mobility. Then, in about 135 BC an Alexandrian mathematician constructed a special kind of clepsydra. The water dripping from the vessel drove a water wheel, which raised a small figure holding a pointer that indicated hour markings along the side. This must have been one of the first forerunners of the numerous mechanical toy clocks designed later.

It would exceed the scope of this book to describe all the other varieties of clock, especially sundials, that were built and used later. This excursion into the field of chronometry and calendar systems is included to show the great importance that the measurement of time and counting of years had attained in Babylon and Egypt. Needless to say, observation of the planets was not neglected in ancient Babylon and Egypt, although little was known about them. An impetus to observe the planets closely was given to Babylonian astronomy about 500 BC by the Persians when they extended astrological prophesies from the conventional scheme of omen astrology to personal predictions and developed horoscopes for individuals. Astrology then became triumphant with the hypothesis of the immortality of the soul and its origin in heaven, and astronomy became its obedient servant.

Elsewhere—in China, India, Persia, virtually all of Asia and the Mediterranean region, as well as in the Central America of the Mayas—development proceeded along similar lines. Unlike Egyptian astronomy, which reached a dead end in its decan theory, Babylonian astronomy fertilized the Hellenistic civilization. Indeed, the era of the Greeks must be regarded as the climax of mathematical astronomy. Whereas the ancient Babylonian astronomy considered events mainly from the perspective of arithmetic, Hellenistic astronomy can be seen in almost geometrical terms. We shall appreciate this even more when we examine the various cosmologies.

The astronomy of the Chinese, which can be traced as far back as the third millenium BC, was, like its Babylonian counterpart, inspired by the need for a calendar. The Chinese, incidentally, also established a lunisolar year and arrived at the same solution as the Babylonians. In the Middle Kingdom eclipses, comets, and other unusual celestial events were observed in great detail. The Chinese believed that heaven reflected events on earth. The court astronomers were also the monitors of all events and were appointed to interpret the influence of these phenomena upon the state. How seriously this was taken is shown by the story of two court astronomers, Hi and Ho, who failed to predict a solar eclipse and were beheaded for their negligence. Historical and astronomic research established that this solar eclipse occurred on October 22, 2137 BC.

Far from the European civilizations, the Mayas of Central America developed an evidently highly advanced astronomy; unfortunately little of it has come down to us. All we know is that the Mayas observed the motions of the moon and of the planets and had highly accurate information about their times of revolution, about the path of the moon, and about other events. Astrological elements, too, played a minor part. The desire for a well-functioning calendar was central to their research. The Mayas had a calendar in which several independent periods ran side by side. The point from which counting began is not known with absolute certainty; according to one interpretation the Maya calendar began on November 11, 3373 BC; this date indicates a very early and enduring interest in the courses of the stars.

A PICTURE OF THE WORLD

Let us at last examine the question of the development of an astronomic picture of the world. We do not know when man first questioned the design of the terrestrial and of the celestial world because we have no historical evidence of that very early epoch. Be that as it may, the ancient ideas about the world were mythical; they did not include observed facts. A Babylonian epic poem from about 2150 BC is an example. It speaks of three gods who destroyed the "primeval gods of chaos" and as a reward were moved to the sky as constellations of the zodiac. This was an astronomic reference, but the result was a mystical picture of the world without any relation to observations. The fact that in Babylonian civilization no speculation independent of mysticism, superstition, and religion, let alone observation about the design of the world, existed is connected with the form of astronomy pursued in Babylon: it was the exclusive preserve of the priests. Nobody outside the religious caste bothered to ask questions about the motions of the stars, their nature and their origin. Of necessity, then, all cosmological considerations were a matter for the priests and therefore suffered from the prejudices inherent in religious, or at least in mythological, attitudes. The origin of astronomy must clearly be sought in

Babylon; that of the uncommitted cosmology must be sought in Greece. The activities of the priests of Babylon can at best be described as mythological cosmogony.

Aside from the already mentioned ideas about the structure of the earth and sky, the Babylonians gradually evolved various other concepts. In one theory the earth was thought to be a mountain. According to other theories water was the origin of all matter; hence the world inhabited by man originated in "the depth" and was surrounded by the ocean, on whose distant shores the sun god put his cattle to grass. Later it was assumed that the sky surrounding the earth was a solid vault whose foundations rested on the ocean, in "the depth," which was also the foundation of the earth. Above the vault of the sky were the "upper waters," and above these "inner heaven"—the abode of the gods. There was the "sun house" from whose eastern door the sun emerged in the hour of the morning, to disappear through the western door in the evening. Ideas similar to those of the Babylonians (unfortunately, too many to mention here) are also found in the Old Testament, without any definitive statements about the shape of the earth. Only the "circle of heavens" and the "pillars" or "foundations" supporting the earth are repeatedly mentioned.

Comparable ideas about the structure of the earth and of the universe were entertained by the Egyptians, who imagined the universe to be a large, rectangular box, with the earth serving as the bottom. The longitudinal axis of the box corresponded to the north-south direction, the direction in which the country of Egypt extends. The surface of the earth was regarded as a narrow, slightly raised disk with Egypt as the center. The Egyptians held the view that in the beginning there was water in eternal darkness, and the world was created out of the spirit of the water. Presumably, they adopted this idea from the Asians who emigrated from Babylon into Egypt.

The idea of water as the origin of all matter is also found among the earliest mythological speculations of the Greeks about the structure and cosmogony of the world. But this idea probably evolved independently and was not adopted from the Egyptians.

The earliest indications of the ancient Greek ideas of the earth and the universe have come down to us in Homer's epics, dating from the eighth century BC. At this time the earth was thought to be a flat, circular disk surrounded by an ocean which began at the Pillars of Hercules (Gibraltar), turned northward, eastward, and southward where it flowed back into itself. A few centuries after

Homer, Greek philosophy began to develop further. As it became established, preoccupation with the questions of the structure and the history of the earth and the sky was independent and far more intense. Many ideas were evolved which, in the course of time, were increasingly based on observation and sought to explain visible facts. Fewer and fewer were pure speculation.

We will consider only a few cardinal points of this development without entering into the details and arguments of the various schools of philosophy. It is remarkable that as early as 350 BC philosophers taught that the earth was round, a concept later forgotten for centuries. This theory of a spherical earth can be traced back to the school of Pythagoras (570–497? BC) and has its origin in the observation of nature. The Pythagoreans adduced as evidence of their theory that when a ship appears on the horizon first its mast comes into view, followed by the lower structures as it approaches the land, and finally the whole ship is seen. It gradually climbs the "water mountain" between it and the viewer resulting from the spherical shape of the earth. The Pythagoreans also pointed to the circular shape of the earth's shadow during lunar eclipses. Their interpretation of this natural phenomenon hit the nail on the head with the assumption that during a lunar eclipse the moon enters the shadow of the earth. In addition they had philosophical reasons ready: the school of Pythagoras considered the sphere the most perfect of forms. What other form could the earth have than that of the sphere? About the position of the earth in the universe they thought at first that the earth resided in the center of the universe, but later taught, with a philosophical explanation, that the center was occupied by the "central fire" around which the earth revolved (fire was considered a nobler substance than that of which the earth was made). Obviously, the noblest substance had to occupy the center.

Anaxagoras (500–428 BC), a contemporary of Pythagoras, developed a model of the world which although still far removed from reality was on the whole free from inherent contradictions. According to Anaxagoras the world consisted of the substances ether and air—primeval matter which at the beginning of the universe was made to rotate "by the spirit." The air, because of its gravity, then moved toward the center and finally developed through the intermediate stages of water and mud into solid rock. The moon was described by Anaxagoras as earthlike; he evolved ideas about its size and postulated the existence of mountains and valleys on it. He explained its phases and eclipses, noted that it reflected the light of the sun, and made assumptions about the distances between moon, sun, and planets. But his views were soon forgotten, particularly since they earned him the accusation of godlessness, and he was able to have his sentence of death commuted into exile only through the good offices of a friend.

The only one of his ideas to survive was that of the earth being a globe. Plato (427–347 BC), adopted the theory of a spherical earth from the Pythagoreans, although the Platonists were not prepared to accept the ideas of the central fire, or an earth that was not the center. According to them the earth occupied the center of the universe, and the sphere of the stars revolved around it.

At that time, however, attempts were made to harmonize theory and observation. Eudoxos (409–356 BC), a Platonist, tried to explain the motions of the planets. See box, "Stations, Retrogressions, and Loops," for a more detailed elucidation of this phenomenon.

As did Plato, Eudoxos believed the earth to be the stationary center of the universe. In order to explain the strange motions of the planets around this assumed center he developed a system of concentric crystal spheres whose interaction was able to shed light on the strange retrogressions of the planets. Since all the works of this great mathematician have been lost, Eudoxos' theory has not come down to us in its original form. But the theory of the motions of the planets and of the concentric spheres which caused them was included in detail in a history of astronomy by Eudemos of Rhodes in the fourth century BC. The Greek philosopher and astronomer Sosigenes then copied it in one of his own works during the second century AD. Unfortunately, both works have been lost. However, Simplikios, a commentator of Aristotle, described it in his book *De Coelo* (*The Sky*), therefore we know the details of Eudoxos' model. Obviously, to make theory fit observation, Eudoxos was obliged to introduce more and more concentric spheres, and his system, although it was most elegant mathematically, became complex and unwieldy. He regarded this explanation merely as a mathematical model; Aristotle (384–322 BC), however, thought the spheres were real, to him the model was reality.

Aristotle also applied himself to the then topical question of the size of the (spherical) earth. He quotes an estimate but does not mention his source. At any rate he already knew that on journeys to the north and south different stars appear in the zenith and rise

Stations, Retrogressions, and Loops

A few weeks' observation of the night sky shows that the planets (from the Greek for "wanderers") move relative to the fixed stars. But Mercury and Venus can be observed only shortly before sunrise and shortly after sunset. Their position relative to the sun changes very quickly. They are the "sprinters" among the planets. Mars moves a little more slowly, Jupiter and Saturn slowest. These were the five planets known to the ancients. The sixth, Uranus, although it can occasionally be seen with the naked eye, is so faint that it is quite inconspicuous in the ocean of stars. Neptune and Pluto can be observed only through powerful telescopes, and then only with great difficulty.

Planetary motion normally proceeds against the diurnal revolution of the sky (from west to east), just as the moon, and the sun (a little more difficult to prove from observation) do move from west to east or right to left among the stars in their annual paths. The time required by a planet to return to a certain star (or, more precisely, to complete one circle of longitude) after one orbit around the sky is called the sidereal period. Mercury requires 87.969 days, Venus 224.701 days, Mars 686.980 days (1.88 years). The sidereal period of Jupiter is 11.86 years, that of Saturn 29.46 years. But this motion is not always what is called "direct motion"; the planet sometimes seems to come to a sudden stop—a station or stationary point—on its path. Then it apparently moves backward—retrogression. After this the planet stops again briefly—the next station—and resumes its direct motion. It may often double, loop, and curve. Today we know that these apparent movements are caused by the relative movements of the planets and the earth around the sun. The retrograde movement, for instance, appears when the earth overtakes another planet or is being overtaken in its orbit. The loops are caused by the different inclinations between the earth's orbit and the orbital planes of the other planets. It took man a long time to find this out—a stationary earth was assumed for the larger part of history.

The illustration shows such a retrograde motion with the two stations of the planet Mars during 1979 and 1980. In October 1979, Mars was in the constellation Leo and proceeding along the direct motion. On January 17, 1980, it stopped and then reversed (retrogression); after a second station, on April 7, the planet continued along its direct motion. These motions can be similarly observed with Mercury and Venus, Jupiter and Saturn.

above the horizon (on southward journeys) and disappear below it (on northward journeys). Because this phenomenon was noticed even on short trips he concluded that the earth could not be all that large. He then quoted (again, without mentioning his sources) mathematicians who attempted to determine the earth's circumference and arrived at a value of 400,000 stadia. From this he drew the further conclusion that compared with other stars the earth is not very large. We do not know the exact length of the Greek stadion. If we base it on the most probable value, 157.5m, we arrive at a circumference of 63,000km[2]. at any rate of the same order of magnitude as the true value of 40,077km.

Orbit of Mars from October 1979 (point 10) to July 1980 (point 7).

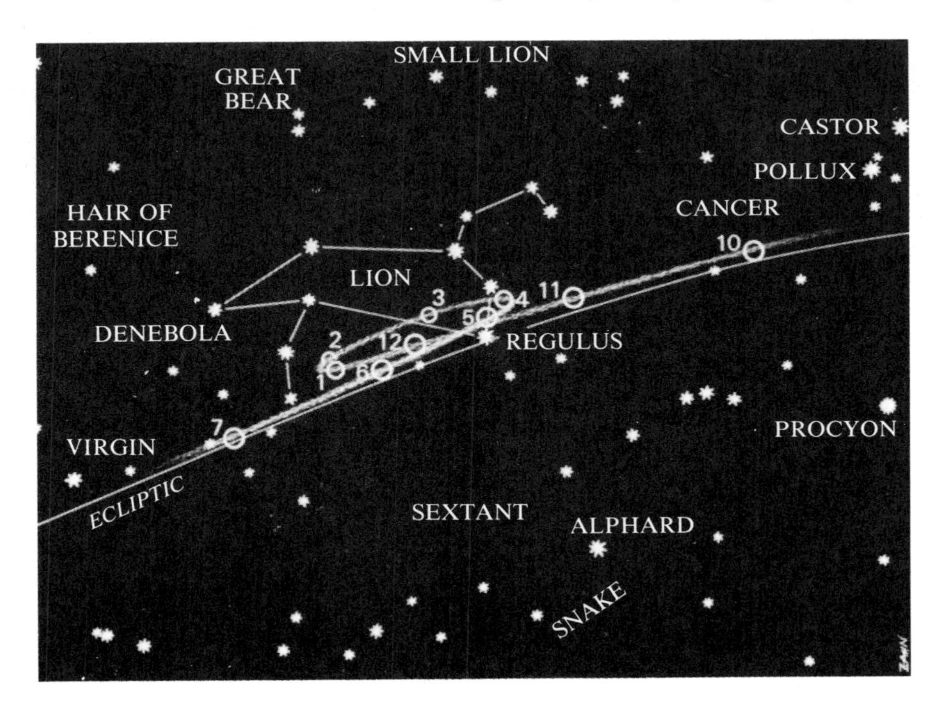

The figures represent the position of Mars at the beginning of each month. For thousands of years these looping motions could not be explained. Ptolemy introduced the epicyclic theory to try to explain the motion. Not until the sixteenth and seventeenth centuries was the true explanation found. Copernicus and Kepler discovered that these motions are the result of the planets' irregular, elliptical orbits around the sun.

As far back as the first and second centuries BC the Greeks had established reasonably correct ideas about the size of the earth, the moon, and the sun. The dimensions of the planets were still a mystery, and it was not until the invention of the telescope at the beginning of the seventeenth century that accurate knowledge was gained. The picture shows the largest planet, Jupiter, compared in size with the earth. Note also the size of the shadow cast; on its clouds, by one of Jupiter's moons.

Eratosthenes of Kyrenia (276–194 BC) carried out the first known measurement of the earth. He was the librarian at Alexandria and had heard that the sun at Syene, south of Alexandria, did not cast any shadow at all at midday of summer solstice day. In Alexandria it cast a very distinct shadow. The task was simple arithmetic. All Eratosthenes had to do was measure the distance between Syene and Alexandria and the length of the shadow the sun cast in Alexandria on summer solstice day at noontime to be able to calculate the circumference of the earth.

Eratosthenes is best known as a geographer. He is said to have written a book entitled *Libri Dimensionum* (*Books of Dimensions*) about the calculation of the earth's size. As this book, also, has been lost, we know the result of Eratosthenes' considerations but not all the details. It is, for instance, occasionally mentioned that Eratosthenes measured the distance between Syene and Alexandria by riding in a carriage and counting the revolutions of a wheel. According to other (and more reliable) sources the distance of 5,000 stadia had already been determined and was simply used as a basis by Erathosthenes. Neither is it certain whether he himself observed and measured the sun's altitude in Alexandria on midday of summer solstice day or already knew the value from hearsay. He was at any rate aware that the sun was one fiftieth of the entire circumference away from the zenith (the point vertically above our heads). That it was in the zenith on summer solstice day at Syene, Eratosthenes had gathered from information that on this day the sun's rays reached the bottom of a well. This meant that at that moment the sun did not cast any shadow and was indeed vertically above the people's heads in the zenith.

On the basis of this knowledge Eratosthenes was able to calculate the earth's circumference by multiplying 50 × 5,000—50 because the distance between Syene and Alexandria must be one fiftieth of the earth's circumference; 5,000 because this was the distance in stadia between the two cities.

As we now know, the calculation suffered from several errors. Syene, the modern Assuan, is not due south of Alexandria, neither does the sun there reach precise zenith at noon on summer solstice day. Furthermore, the 5,000 stadia was most probably only a round value. The geographer Strabo (app. 63 BC–AD 26) claims (in his extant works) that according to Eratosthenes the distance between the smallest cataract at Syene and the Mediterranean Sea near Alexandria was 5,300 stadia. All these minor errors seem to have cancelled each other by sheer coincidence. It is true that we do not know for certain the value Eratosthenes assumed for the stadion, but if we base it on 157.5m, the value Pliny used, we arrive at a circumference of the earth of 39,681km and a diameter of 12,631km. This approximates the correct measurements known today to as high an accuracy as 1 percent.

Preoccupation with the size of the earth, the moon, and the sun was quite widespread at the time, and Eratosthenes' attempt and method to determine the size of the globe were by no means unusual. Aristarchos of Samos (310–230 BC) wrestled with the same task. His ideas have come down to us in a work by Archimedes. They culminate in the statement that the sun is the center of the universe and that the earth, moon, planets, and stars revolve around it in concentric orbits.

Aristarchos of Samos was the first scholar to postulate the heliocentric system. The clearest transcription of Aristarchos' ideas come from Plutarch's (app. AD 50–124) *De Facie in Orbe Lunae* (*The Face in the Moon's Disk*) written more than 300 years after Aristarchos' work. In it Plutarch, who obviously owned a copy of Eratosthenes' *Libri Dimensionum,* has one of his contestants say:

> Do not go, my dear friend, and lay an accusation against me of godlessness after the manner of Cleanthes, who thought it was the duty of the Greeks to accuse Aristarchos of Samos of godlessness because he moved the focus of the universe on the basis of his attempt to save the phenomena by assuming that the sky was motionless and the Earth revolved in an oblique circle while at the same time rotating round its own axis.

Here, then, we find not only the indication of the motion of the earth around the sun but also of its rotation on its own axis. The mention of the oblique circle may refer to the fact that the earth's equator and thereby its axis of rotation is inclined to the orbit the earth describes around the sun. This explains the origin of the seasons. The passage "to save the phenomena" was, in fact, a well-known mode of expression the Greeks used when they wanted to reconcile theory with reality.

The whole problem of Greek astronomy, as Plato noted, consisted in the effort to prove the philosophical postulate of the uniformity of all motions in the universe. The circular motion at uniform velocity was, according to the philosophical interpretations, the perfect principle and hence the only one appropriate to heavenly events. However, the observed motions of the planets were irregular both in velocity and in the shape of the orbits. For Greek philosophers the problem reified the task of representing those irregular motions of the planets by uniform motions in circles.

The experiments of Eudoxos and Aristotle with the crystal spheres (of which all of twenty-seven were required even for a shaky explanation of all the motions in the planetary system) was just one of several attempts, but it constituted at any rate a hypothesis that was to dominate human thought for many centuries. Although his heliocentric theory enabled Aristarchos of Samos to dispense with the crystal spheres and still explain the loops of the planets, he could not account for the irregularities of the motions of the planets; because he, too, proceeded from regular motions in circular orbits. In spite of its acceptance—there was still a school of Aristarchos a century later—the heliocentric theory was not outstandingly successful.

Seemingly objective arguments against the idea of an earth revolving around the sun joined forces with philosophical prejudices to oust the heliocentric theory. A frequent objection was that, were the earth to revolve around the sun, the stars would have to exhibit parallax displacement in the rhythm of the year. A mental experiment will show what this means: if we sight an object that is not very far away from our eyes (say, a candle) with only one eye and the other eye closed, and immediately afterward close the open eye and open the other, the candle appears to be displaced against the background. The situation would be the same, so the argument ran, with the stars at half-yearly intervals, were the earth to revolve around the sun in a circular orbit. As an ''image'' of this orbit, the stars would also have to describe small circles in the sky. Aristarchos, fully aware of this objection, had refuted it with the claim that the stars were too far away, so the radius of the sphere of the stars was much too large to make observation of this phenomenon possible. Although his argument was absolutely correct, it carried little conviction.

Further objections against the heliocentric system and a revolving earth were founded on purely philosophical prejudices. Man, the Measure of All Things, preferred his earth to be the center of the universe rather than in any odd, nonprivileged place. Added to this was the assumption that on a revolving earth all objects would have to lag behind this revolution, there would have to be a universal air stream. These arguments heralded a return to the geocentric system, especially as propagated by the leading philosophers of the next centuries. Apollonios, Hipparchos, and Ptolemy, the great Greek scholars from the second century BC to the second century AD, were keen protagonists and developers of the geocentric system.

COLOR CAPTIONS

(Page 27, top)
This view of earth was taken during the voyage of Apollo 17 to the moon. The picture shows the area from the Mediterranean to the Antarctic and the South polar ice-cap from a distance of about 240,000km.

(Page 27, bottom)
This infrared photograph of the Salton Sea and the Imperial Valley in southern California was taken during the flight of Apollo 9. The red quadrangles are cultivated fields. The Colorado River and part of Arizona appear on the right of the photograph; the towns of El Centro, California, and Mexicali, Mexico, are at the bottom.

(Page 28, top)
This is a photomicrograph, in polarized light, of lunar rock obtained during the Apollo 17 voyage. Such pictures provide information about the mineralogical structure of the rock.

(Page 28, bottom)
This is the scientific research station set up on the moon during the Apollo 17 exploration. The central station (center) transmits measured data. Also visible are a mortar that fires grenades to produce artificial moonquakes (left) and the equipment for measuring charged particles in the vicinity of the moon (right).

(Page 29)
Morning mist above the surface of Mars.

(Page 30, top)
Jupiter from a distance of 12 million kilometers. Io, the innermost of the Galilean satellites, can be seen above the turbulent layers of cloud. (Photograph taken by Voyager 2, June 25, 1979.)

(Page 30, bottom)
The Great Red Spot, from a distance of 6 million kilometers, and the huge turbulences frequently produced in the Jovian atmosphere. (Photograph taken by Voyager 2, July 3, 1979.)

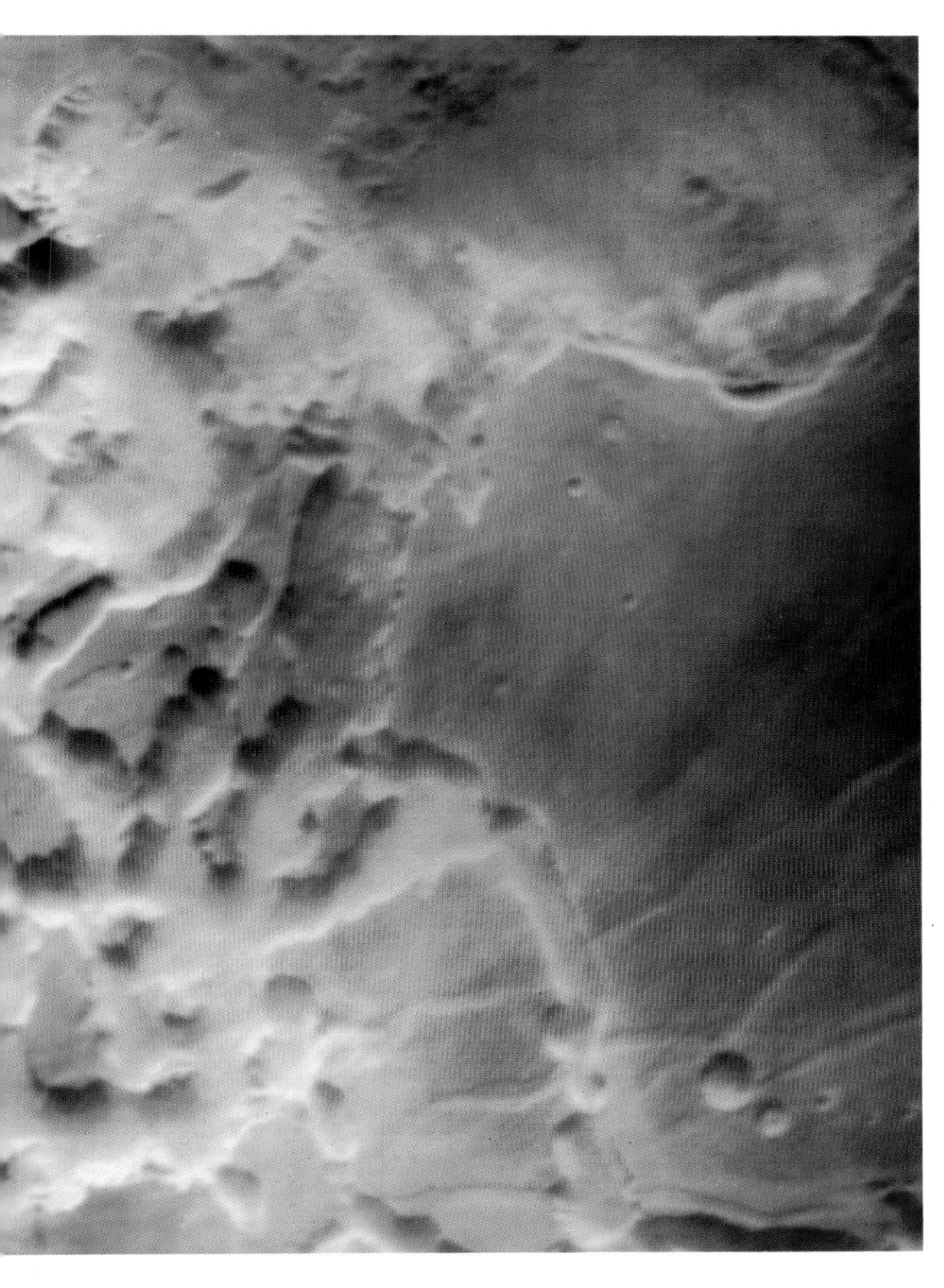

IO

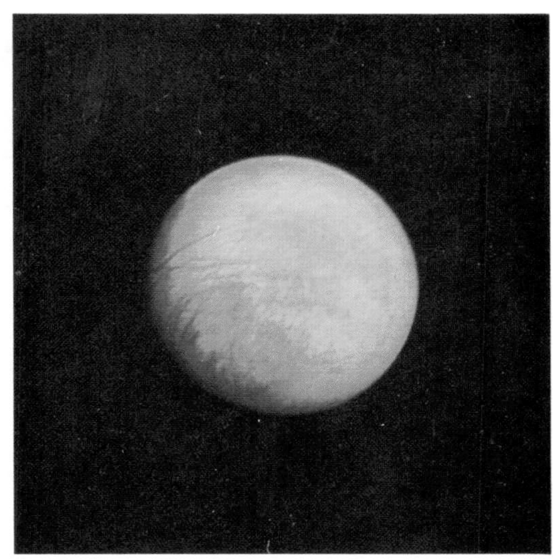

EUROPA

GANYMEDE

CALLISTO

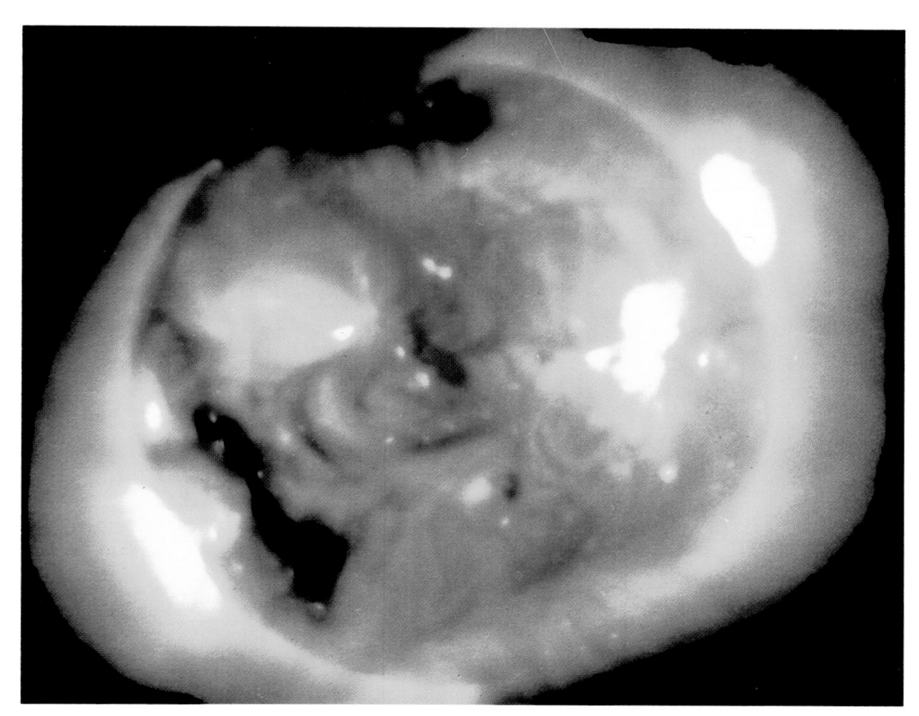

(Page 31)
The Galilean satellites, photographed by Voyager 1. They are shown in correct scale. Europa is about the same size as the earth's moon.

(Page 32, top)
This picture—taken by Pioneer 11 on September 1, 1979, from a distance of 395,000km—shows the division between Saturn's rings and the low density of the ring material.

(Page 32, bottom)
The details in the atmospheric cloud aspects of Saturn are clearly shown in this picture taken on October 18, 1980, from a distance of about 25 million kilometers.

(Page 33, top)
This picture of the sun in x-ray light was obtained by Skylab. It shows part of the corona within the temperature range of about 1,000,000°C. The dark areas are "corona holes"—areas at lower temperatures—which are vents of the solar wind.

(Page 33, bottom)
False-color photograph of a solar eruption in the extreme ultraviolet range of the spectrum, taken by Skylab.

(Page 34, top)
A recent picture of the Crab Nebula in the constellation Taurus. The nebula is the result of an eruption of a supernova in 1054. In the center of this structure, at a distance of 6,300 light years and still expanding, is a neutron star which is also a pulsar in the radio region and in visible light.

(Page 34, bottom)
The Andromeda Nebula, our nearest extragalactic star system, 2.3 million light years away.

The Ptolemaic System

Claudius Ptolemy (about 100–160 AD) perfected and described in detail the geocentric system, which still bears his name. It has come down to us in his *Mathematical Collection*, a work generally known today as *Almagest*. It stands to reason that the Ptolemaic system was not developed by Claudius Ptolemy alone. It contains elements of the work of the mathematician and astronomer Apollonios of Perge (third century BC) and the astronomer and geographer Hipparchos (app. 180–125 BC). In the four hundred years between Aristarchos and Ptolemy, these were the only scientists to study the theory of the motions of the planets. They were also conscientious observers and recorded much data about the motions of the stars and thus contributed to the increasingly scientific status of astronomy. Their aspirations were not necessarily directed toward finding the true explanation of the motions of the heavenly bodies; what they hoped to establish was a method of describing and predicting the observed motions of the stars.

The pillars of the Ptolemaic system are the epicyclic theory and the eccentric motions of the planets. The epicyclic theory was evolved by Apollonios, and its beginnings may date back even further; Hipparchos developed the eccentric motions of the planets. Let us examine how these elements are represented in the Ptolemaic system.

Ptolemy assumed that the planets revolve in epicycles around the globe of the stationary earth. A circle, the *deferent*, was imagined around the earth. The center of the deferent was not in earth, but displaced a little outside it. In other words, the earth was not precisely in the center, but eccentric. A point that formed the center of a second, smaller circle, the *epicycle*, revolved on the deferent. This arrangement was repeated several times; the center of a second epicycle moved on the first epicycle and so on, until eventually a planet moved on the last epicycle.

This was the only way—with the aid of eccentric deferents and epicycles—to arrive at a rough-and-ready explanation of the motions of the planets, especially stations and retrogressions. This is also how a mathematical-geometric structure of a nest of circles evolved. They were necessary merely to explain the observations. There was no physical justification why a planet should move on a circle that in turn moves on a circle.

These theories were not the only astronomical problems and items of information to which Ptolemy addressed himself in the *Almagest*. Among many others was consideration of the Star Catalog compiled by Hipparchos almost three centuries earlier. It contains twenty-one constellations, comprising 1,022 stars in six brightness classes or orders of magnitude as we say today. It also includes descriptions of solar and lunar eclipses, ancillary tables, etc. In *Hypotheses planetarum* (*Hypotheses About the Planets*), Ptolemy even attempted calculations of the size of the universe and of the bodies of the stars and arrives at a distance of about 130 million km between us and the outermost sphere of the stars.

For centuries the Ptolemaic system was accepted as the valid picture of the universe. Of course there were those who proposed other opinions about the structure of the universe, the large number of inhabited worlds, the rotating earth, and so on. Initially tolerated by the spreading Christian Church, these dissenting opinions were later increasingly attacked. The Church took the side of Aristotle and regarded the Ptolemaic system as a welcome method of

In 1428 the Tatar prince Ulugh Beg built this observatory at Samarkand. Now in ruins, the picture shows how it must have looked in the fifteenth century. Ulugh Beg determined the positions of the stars with quadrants, and in 1437 he corrected the Ptolemaic Star Catalog. The Catalog was probably compiled by Hipparchus during the second century BC and was the first product of astronomical observation.

calculating the motions of the stars, mainly from the aspect of the calendar. It can, indeed, be seen that with the advance of Catholicism the cosmological picture of the world became more and more superficial and confined. During the Middle Ages scholars showed little interest in the calculations of distances that had engaged the minds of Greeks at about the time of the birth of Christ. The hypotheses of these "heathen philosophers" were rejected, indeed suppressed. It was feared that the attempt at philosophical and scientific "explanations" of objects would diminish the greatness of the Divine Creator. Holy Writ was regarded as solely binding. The concept of the spherical shape of the earth was denied, maligned, and replaced by that of a flat earth in a narrow universe. Everything that contradicted this very odd picture of the world was declared heresy and heavily punished.

This state of affairs changed only centuries later under the gradual impact of fresh knowledge. Arab influences contributed to this development from the seventh century AD onward, established by the Arab conquest of Spain and North Africa. The Arabs sought to suppress both Greek philosophy and Christianity, because contemporary Arab civilization was orientated toward and interested in astrology.

The New Cosmology

Not until the ninth century did the idea of the earth as a globe regain ground, and Greek thought resurfaced in Latin translations. The Church sometimes attacked and sometimes tolerated these seeds of a new cosmology. We shall lose no time in describing the breakthrough to the new picture of the universe, linked to the names of Nicolaus Copernicus, Galileo Galilei, Johannes Kepler, and Sir Isaac Newton.

These representations of the Ptolemaic and Copernican systems were published by Cellarius in 1708. Ptolemy (opposite, top) shows the earth in the center of the universe, circled by the moon, Mercury, Venus, the sun, Mars, Jupiter, and Saturn. Copernicus (opposite, bottom) places the sun in the center, orbited by Mercury, Venus, Earth and its moon, Mars, Jupiter and the Galilean satellites, and Saturn.

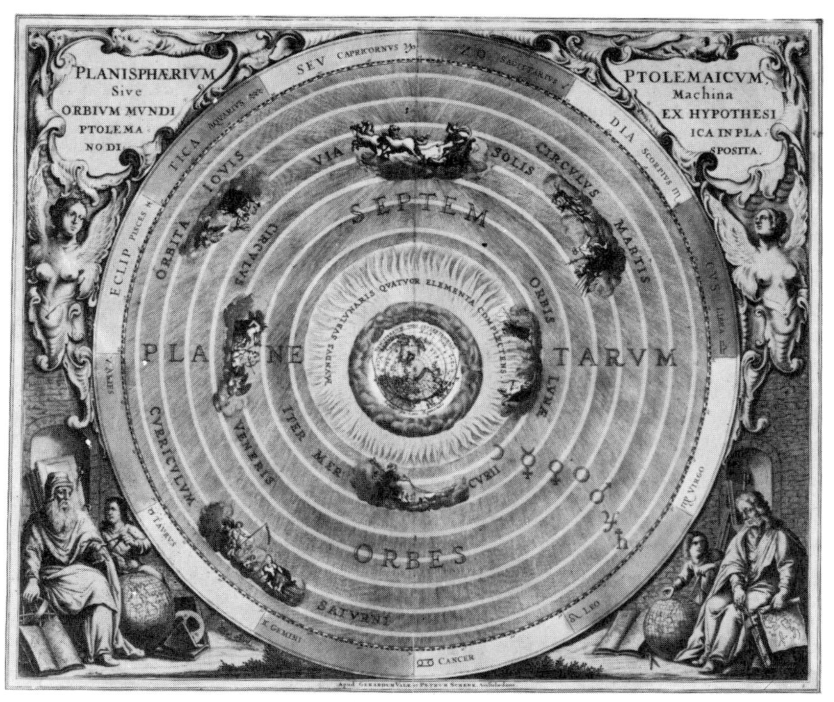
PLANISPHÆRIVM
Sive
ORBIVM MVNDI
PTOLEMA
NODI
PTOLEMAICVM,
Machina
EX HYPOTHESI
ICA IN PLA
SPOSITA.
SEPTEM
PLA NE TARVM
ORBES
ORBITA IOVIS
VIA SOLIS
CIRCVLVS MARTIS
ORBIS LVNÆ
ITER MERCVRII
SATVRNI
CAPRICORNVS
SAGITTARIVS
SCORPIVS
PISCES
CANCER
LEO
VIRGO
GEMINI
TAVRVS
MVNDVS SVBLVNARIS QVATVOR ELEMENTA COMPLECTENS

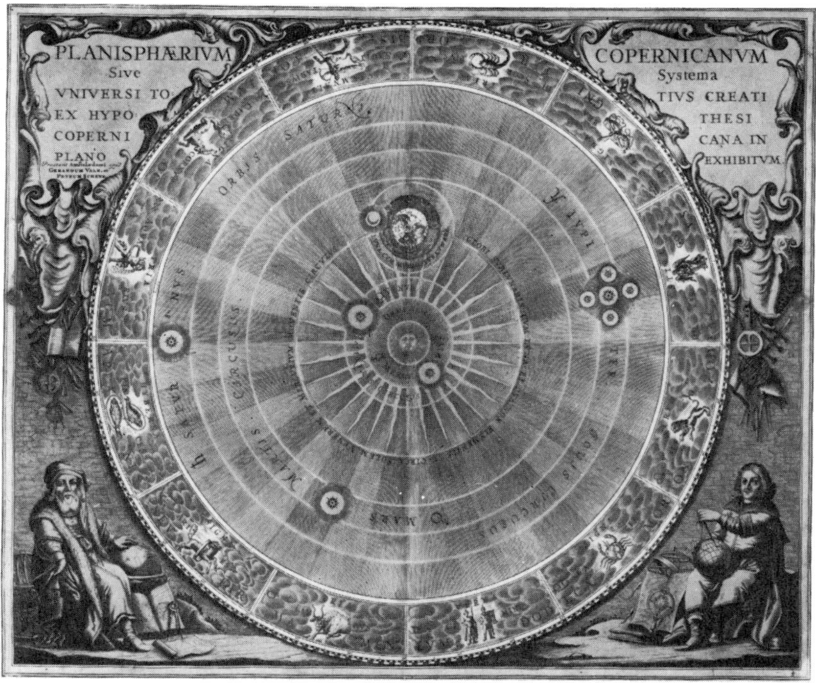
PLANISPHÆRIVM
Sive
VNIVERSI TO-
EX HYPO
COPERNI
PLANO
COPERNICANVM
Systema
TIVS CREATI
THESI
CANA IN
EXHIBITVM
ORBIS SATVRNI
MARS

In 1543, the year of his death, the Frauenburg canon Nicolaus Copernicus published his famous work, *De Revolutionibus Orbium Coelestium* (*The Revolutions of the Heavenly Bodies*), in which he hypothesized that it is not the earth, but the sun that occupies the center of the world, and that the spherical earth, rotating around its own axis, revolved around the sun. Copernicus had arrived at this view after years of deep conflict following the study of many publications of antiquity. He had found contradictions in the Ptolemaic system, deviations in the star catalogs of earlier periods, and different methods of treatment of the orbital deviations of the planets. He then began to study the works of the Greek philosophers, became familiar with the idea of the rotating earth, and Aristarchos' heliocentric system.

These theories and hypotheses at first appeared odd to him. But he made use of earlier observations as well as contemporary ones. He did not merely look for a system that described the motions of the stars better than the systems it would replace. What he wanted to find was the truth; and this, so he argued, could be proved by the facts established by the observation of the stars. As early as 1510 he came to the realization that the world must be designed heliocentrically, that the sun must be the center of the universe. He spent the following two decades on underpinning his heliocentric system. In doing this he failed in what he had hoped to be able to do, that is, to eliminate the epicycles. Even Copernicus could not perfectly describe the motions of the planets on the basis of regular circular heliocentric orbits.

With the heliocentric idea Copernicus had grasped the truth, but not the whole truth. Only a short last step remained to be taken: if Copernicus had accepted an irregular motion of his planets and replaced his circles with ellipses, he would have found a picture of the universe that agreed with reality. But the great astronomer was unable to take this last step; his basic philosophical outlook prevented this. Copernicus was a neo-Platonist. To him the form of the circle and regular motion were perfect, divinely inspired properties in the realm of the stars, and he was unable to abandon this outlook.

Another 70 years were to pass before the successful breakthrough of a new philosophy of cosmic motions. These years were racked with controversy over the Copernican theory, a controversy that reached its climax when Johannes Kepler and Galileo Galilei took the next step in finding the truth about the nature of the universe.

The mathematician and astronomer Johannes Kepler (1571-1630) had been attracted by astronomy in his youth, although he had intended to become a Lutheran pastor. But during his advanced theological studies at Tuebingen University he also attended lectures on astronomy and enthusiastically accepted the Copernican system, although his teacher Maestlin taught and advocated the Ptolemaic one to avoid the risk of losing his professorship.

On Maestlin's advice Kepler accepted the post of mathematics master at a school in the Austrian city of Graz; but in his spare time he practiced astronomy. He looked for links between and laws governing the distances of the planets from the sun and thought he had found them in the five regular geometric solids (from the triangular pyramid with three coincident faces to the twenty-face icosahedron). The outcome was his first book, *Mysterium cosmographicum* (*The Cosmographic Mysteries*). This work brought him into close contact with Tycho Brahe, perhaps the most outstanding astronomical observer of the time; Kepler eventually became his assistant at the court of Emperor Rudolph II in Prague. After Brahe died in 1601 Kepler succeeded him as Imperial Mathematician and Astronomer. He adopted thousands of Brahe's astronomical observations, among them sightings of the planet Mars, which he used to calculate its orbit. But the calculations refused to add up. After years of tedious calculating and reasoning Kepler realized that the error consisted in Copernicus' assumption that the planets described regular circles round the sun: they moved in ellipses. He continued his calculations of the orbit of Mars and found a second important fact. It concerned the velocity of the planets: they move faster when at the point nearest the sun, the perihelion, than when at the point farthest away, the aphelion. These discoveries are today known as Kepler's First and Second Law:

1. The planets move in elliptical paths with the sun in one of the foci.
2. The radius vectors of the planets describe equal areas of the ellipse in equal times.

Kepler published these laws in 1609 in a book dedicated to Emperor Rudolph, now known as *Astronomia Nova* (*The New Astronomy*). The original title was much more long-winded: "The New Astronomy, or Celestial Mechanics, Its Reasons Explained with Commentaries on the Motions of the Star Mars, According to the Observations of Tycho Brahe." In 1619 Kepler included in *Harmonice Mundi* (*The Harmony of the World*) a third law; this established a link between the periods of revolution of the planets around the sun and their distances from it:

3. The squares of the periods of revolution of the planets around the sun are proportional to the cubes of their mean distances from it.

This brief discussion does not do justice to the achievements of Johannes Kepler. With the formulation of the three laws bearing his name he not only created the possibility of at last being able to calculate the motion of the planets precisely. What is perhaps more important is in doing so he overcame an ancient mental prejudice (the theory that the planets move in circular orbits at a constant velocity) which blocked a breakthrough of the Copernican idea of the heliocentric system of the universe. In addition, Kepler followed the principle, by no means generally introduced at the time, of basing his research of the findings of his observations. This placed him on the threshold of a new era of the natural sciences.

The same applies to his contemporary, Galileo Galilei (1564–1642). This Italian physicist became famous both for the formulation of the laws of gravitation and as the first man to point a telescope at the star-filled sky. He thereby introduced an instrumental aid into astronomy whose importance can hardly be underestimated. In March 1610, Galileo published *Siderius Nuntius (The Messenger of the Stars),* in which he wrote:

> About 10 months ago a rumour came to my ears that a Belgian had made a glass for looking through, with which objects, although far away from the eye of the observer, would clearly appear as if they were near when seen through this glass, and a few examples were given which were believed by some and doubted by others. This gave me the decisive impulse to direct all my thoughts towards investigating the causes and finding the means of inventing a similar instrument, and guided by the laws of dioptrics I soon succeeded. I started with making a tube out of lead, to whose ends I attached two glass lenses, both plane on one side and one spherically convex, and the other concave on the other side. When I moved the eye towards the concave glass I saw the objects very large and close, they appeared three times closer and nine times larger than when seen by the natural eye. (Freiesleben 1956)

These sentences clearly show that, contrary to repeated claims, Galileo did not "invent" the telescope. On the other hand he did not have a detailed enough description to enable him simply to build a copy. But to Galileo the physicist and mathematician it was nevertheless not difficult to "reinvent" such an instrument within a few hours after he heard of its existence; this is how the telescope he described came into being.

In *Siderius Nuntius (Messenger of the Stars)* Galileo reported in great detail the discoveries he had made with his telescope in 1610. He described how he found mountains and valleys on the moon and how he detected stars one cannot see with the naked eye:

> In addition to the stars of the 6th magnitude (the faintest stars visible to the naked eye), a large number of other stars hidden to the naked eye can be seen with the aid of this instrument; there are so many of them it is almost incredible. In fact, a larger number of them is visible than all the stars of the first six magnitudes taken together. (Ley 1965, p. 129).

Galileo then proceeded to recount in this context his attempt to draw all the stars of the constellation Orion, "but I was overwhelmed by the enormous multitude of the stars and for lack of time I postponed it for a later occasion. . . ." Galileo mentions other star configurations he had drawn, such as the Pleiades, the

The second (top) and third (bottom) telescopes built by Galileo in 1609. The bottom one has an objective of 40.6mm diameter, 8× magnification, and is 92.5cm long. The magnification of the top telescope is 32×; the diameter of its objective, 43.7mm; and the length, 122.5cm.

familiar small group of stars in the constellation Taurus, which, seen through a modern telescope, looks like a flashing diadem. But the scientist was able to record quite different discoveries in *Messenger of the Stars,* discoveries that strongly confirmed his belief in the Copernican system. He found four faint starlets to the left and right of Jupiter—actually, the moons of this planet. Repeated observations at time intervals showed him that they revolved around the giant planet, and he regarded this as proof by analogy of the heliocentric system. As these small moons (Galileo called them "planets" because the plural of the word "moon" did not yet exist) revolved around Jupiter, he argued, so the planets moved around the sun. He also observed that Venus exhibited phases similar to those of the moon: from a large, narrow crescent through first quarter to a small, almost fully illuminated disk. The only possible explanation of this was that Venus, like the earth, revolves around the sun, that the changing phases are due to the changing relative positions of the sun, Venus, and Earth, and that the different apparent sizes are the result of the greatly differing distances from the earth.

Galileo had accepted the Copernican system even before his "re-invention" of the telescope. In a letter (August 4, 1657) to Johannes Kepler, he describes himself as a follower of Copernicus "for many years." But his own discoveries with the telescope were essential to adding the necessary weight to the Copernican theory of the helio-centric structure of the universe. Kepler and Galileo indeed con-summated the entry into the Copernican era. Once more the battle between Ptolemaists and Copernicans flared up, not least in the dis-pute between Galileo and the Church, to be finally decided in favor of the heliocentric system, although the Church continued its fight against it for a considerable time to come.

At this point we must mention the man who provided the *explan-ation* of the motions, not only of the planets, but of all the bodies in the universe and on earth, Sir Isaac Newton (1643–1727). In 1687 Newton formulated the Law of Gravitation. Except for a small last step, Kepler had come close to this idea but was unable to com-prehend its full significance. Sir Isaac, on the basis of Kepler's find-ings, noted that every body exerted an inherent force of attraction corresponding to its mass (or proportional to it, as the mathemati-cians say) and inversely proportional to the square of its distance. Translated into everyday language, this means that two bodies whose distance is doubled attract each other only with one-fourth their original force. When the distance is trebled the force is one-ninth, when it is quadrupled one-sixteenth, and so on. Based on this simple statement Newton was able to explain not only the motions of the planets but also the phenomenon of the tides as an effect of the gravitational force between the earth and the moon.

It was only natural that, after Galileo, observational astronomy should make enormous progress; ever bigger and better telescopes were developed. A large number of outstanding observers followed in Galileo's footsteps with their own important discoveries. Needless to say there were many physicists after Newton who made great contributions to the advance of physics. From Kepler, Galileo, and Newton to our own days astronomy has made giant strides. To describe them in detail would require mentioning dozens of new instruments, techniques of observation, and theories as well as hundreds of names. But this book is not a history of astronomy, the aim is to describe the nature of modern astronomy, the fascination that it radiates. Of course, this is not possible without comparisons. The early ideas and theories about the universe, their development from the first primitive pictures of the world to the beginning of the heliocentric aspect on which our concept of the universe is founded, are here to provide the basis for such comparisons.

THE FOUR PHASES OF DEVELOPMENT

When we look back to the beginnings of astronomy we can divide this development into four phases. The beginning of each new phase, however, became the milestone of an epoch whose course was unpredictable and whose consequences could not be assessed.

The first phase was that of the primitive picture of the world, a picture not oriented to facts but the result of mythology and superstition. The second phase was characterized by the first philosophical-astronomical systems of the universe. Its outstanding feature is the Ptolemaic system. The third phase began with the heliocentric system of Copernicus and includes the work of Kepler, Galileo, and Newton. During this phase the natural sciences were established as exact sciences, no longer philosophical but based on observational and experimental tests. To discuss the importance of this phase in all its aspects would exceed the framework of this book. The fourth phase began at the moment man started to send his instruments into space and ultimately prepared to pay visits to other heavenly bodies himself.

The transitions between the various phases are not clear cut; the boundaries are somewhat blurred. Radio astronomy, for instance, which had its practical beginnings after the end of the Second World War (theoretically, it was established as far back as 1931), must undoubtedly be included in the fourth phase. But this does not alter the fact that the development of astronomy can be divided into four phases; from the historical aspect we are still at the beginning of the fourth.

Astronomy is undeniably the oldest of all the sciences. But today it is also a science that encompasses many other disciplines. It has spilled over into mathematics, physics, chemistry, electronics, precision-mechanical engineering, computer technology, high-frequency technology, and nuclear physics. It has also entered geology, mineralogy, meteorology, and—at least in the context of one question—biology, biophysics, and biochemistry.

Nothing excites our imagination so much as this one question of the existence of relatives in the universe. It is really wishful thinking rather than a question. Are there other thinking beings like man in the universe? And how can we establish contact with them?

CHANCES OF OTHER LIFE IN THE UNIVERSE

The answer to the question "Is there life anywhere in the universe except on earth?" is now largely affirmative. But before we can inquire further how highly developed such life might be, whether it has reached the level of intelligence of human life or has risen far above it, we must first investigate the localities in the universe that could offer life—of whatever quality—a home.

All the stars are like our sun, balls of gas of varying sizes whose surface temperatures range between 40,000°C (in some cases 100,000°C) and 2,400°C. Temperatures in their centers are as high as 10,000,000 to 50,000,000°C, and in the special cases of "Red Giants," higher still. Although the experts today have far extended the limiting conditions in which primitive life can exist, they are unanimous in their opinion that suns are not places in which life in any form is possible. Even at the surface temperature of the sun (let alone in its deeper layers) all substances are gaseous and in the state of plasma. This means that the majority of the atoms are stripped of their electrons and therefore move in a charged state. The existence of complex molecular structures, which is essential to life, must in these conditions be absolutely discounted.

The situation is different in the interstellar masses of gas and dust. Here, numerous carbon, nitrogen, and oxygen atoms have been found in addition to the two simplest atoms, hydrogen and helium. This discovery has been due mainly to radio astronomy. The atoms betray their presence not through lines in the visible spectrum but in the radio wavelength range. As early as 1937 two types of molecule that radiate ultraviolet light were discovered in interstellar dust and gas: cyan (CN) and hydrocarbon (CH). The development of radio astronomy and the extension of spectrum analysis to the radio region led in the 1960s to the identification of other molecules and components: carbon monoxide (CO), hydroxyl (OH), even molecules as complex as water (H_2O), ammonia (NH_3), formaldehyde (H_2CO), and methanol (CH_3OH) were found. During the last few years molecules were added which consist of up to nine atoms, such as dimethyl ether ($[CH_3]_2O$) and ethyl alcohol (C_2H_5OH). Today about seventy types of molecule, many organic compounds among them, are known to occur in interplanetary dust. Are they the precursors of living matter?

The famous British astronomer Sir Fred Hoyle, together with the equally eminent astronomer Chandra Wickramasinghe, developed a hypothesis a few years ago that life had indeed its origin in interstellar space, in the gas and dust clouds found there in great numbers. They proposed that about 4 billion years ago the earth was showered with primeval, primitive life germs through a collision with a small protoplanet to which it also owed its oceans and its atmosphere. This hypothesis is by no means generally accepted, but is at least worth considering.

Such primitive nuclei of life obviously cannot develop into higher life in interstellar space where conditions are not suitable. Even the complex molecules originating in space do not last very long in the prevailing environment. High-energy radiation of the most varied kinds and cosmic radiation split them up into their constituent atoms. How long they persist depends on their nature and on the environmental conditions in which they happen to be. The molecules identified so far in the interstellar clouds have lives between 10 and 100,000 years, but this requires a continuous formation of new molecules to replace decayed ones.

The conclusion suggests itself that solid surfaces of planets are essential to life beyond a primitive stage, and this raises another question: how many planetary systems are there in the universe?

This is one of the many questions that cannot be answered by astronomical observation. Planets are dark bodies that radiate only the reflected light of their sun. The suns themselves are too distant for specific observation; and their planets are so close to them that, even if it were at all possible to see them from earth, they would be swamped by the light of their sun. It is true that we have some circumstantial evidence of the existence of planets of other suns; what we do not know is whether there are planets that are as large (or, more accurately, as small) as the earth among them. This circumstantial evidence of planets of other suns consists in perturbations (deviations in orbit) of these suns caused by the attraction of such planets. Such perturbations can, however, be detected only in stars that are not too remote from us, provided the perturbing planet is very massive. The smallest of such demonstrable bodies are in fact several times as massive as the planet Jupiter, whose diameter is about 11 times, volume 1,347 times, and mass 318 times that of the earth.

The question whether planetary systems of stars are exceptional or common phenomena can of course be approached also from other directions. We could, for instance, draw conclusions about the frequency or rarity of planetary systems if we knew how they originate. Many hypotheses exist about their origin. Theories highly popular today regard the formation of planets as a fairly ordinary event, though not with certain types of star.

There are also involved quantitative approaches to this problem; indeed a formula has been developed, by Dr. Frank Drake, that is supposed to determine the number of advanced civilizations, blessed with technological progress, in our galactic system. This formula consists of several multiplicators, each representing a certain condition that must be met for a life-supporting planet to be established, beginning with the percentage of stars that are loners, through the spectral class of the star and its size, to its age, which must be advanced enough to have allowed sufficient time for the evolution of life. All the multiplicands of this formula are speculative. Not a single one (and we have by no means listed all of them) admits

Each of these many stars is a sun like our own. The photograph (taken with the 5m telescope on Mount Palomar) shows Messier 16, an open star cluster in the constellation Serpens. Whether any stars have planets is a question yet to be solved. But it can be assumed that there are inhabited planets in several regions of the universe.

of concrete items of information based on observed facts. The results are fairly random and influenced if the user of the formula is an optimist or a pessimist, if the user subscribes to or rejects certain hypotheses of star development, planet formation, longevity of technological civilizations, and the like.

Radio-astronomical experiments were carried out at considerable cost with the aim of making radio contact with the (utopian) inhabitants of (utopian) planets that (perhaps) exist somewhere (in our vicinity) in the universe. All this in the hope that these civilizations (at this very stage in the history of their evolution) make use of radio waves for the purpose of communication. Even more far-reaching and expensive are projects suggested for a concerted effort to prove the existence of such intelligent creatures. A project called "Cyclops" envisages the erection of 2,500 radio telescopes with aerials having a diameter of 100m. The cost of the Cyclops facilities was estimated as far back as 1971 at $1,200 million (£600 million) annually over a period of thirty years. This did not include the ancillary cost of construction of approach roads, supply depots, and laboratory facilities.

The search for intelligent inhabitants of planets of other stars has already begun on a modest scale. In 1960 and 1961 Dr. Drake used the radio telescope at Green Bank, West Virginia, to look for signals from inhabitants of other planets. This search, was, however, limited to a single wavelength (of hydrogen at 21cm) and to the hypothetical planets of two stars—τ Ceti (Tau in the Whale) and ϵ Eridani (Epsilon in the River)—ten light years away. He found no signals that could have been sent by intelligent creatures.

The question as to whether such cosmic communications can ever be established is largely a matter of faith; there are as many arguments against as for this assumption. We shall confine ourselves to a few basic considerations. It is probable that a large number of the 200 billion suns in our own galactic system have planets. Some of those planets will offer the conditions for the existence and shelter of highly advanced forms of life. We do not know the limits of such vital conditions, because such beings may radically differ from those that populate the earth. Even if we assume that only 1 percent of all stars in our galactic system have planets, and only 1 percent of these could offer a home to highly evolved life, we will arrive at as many as 20 million such planets in our own stellar system. A few hundred of them should be within our own radius of communication.

Although any conversation with the inhabitants of these planets would be very difficult (at an assumed distance of ten light years, twenty years would pass before we could receive an answer to our friendly "Good Morning"), it should nevertheless be possible to record the one-way radio signals of these other beings or to transmit ours to them. Again, this presupposes that these extraterrestrials employ electromagnetic waves for communication.

Man has existed on earth for a few million years. By far the greatest part of this time he spent in the twilight of prehistory. Perhaps 100,000 years ago he began to become aware of himself, 10,000 (or, perhaps, 20,000) years ago he entered history. The recorded or written historical tradition began a few thousand years ago. And for little more than half a century man has made use of electromagnetic waves as a means of communication. It is almost impossible to say that at exactly the same time in the history of the universe (which according to the latest information has been in existence for about 20 billion years) a group of intelligent beings has reached the same stage of development as we have and uses the same vehicle of communication.

Perhaps there are intelligent beings in our immediate neighborhood in space. Perhaps they sent radio signals into space for centuries or thousands of years and stopped a few years or decades before we discovered electromagnetic waves and used them for communication. Perhaps they will start in 500 or 1,000 years from now. And perhaps (if humanity survives that long) we might then use a completely different medium of communication.

Perhaps these intelligent beings somewhere in the universe, quite differently designed, have quite different interests? Perhaps they have a completely different system of communication technology. Perhaps they are totally uninterested in technology. Perhaps they are not interested in the universe and in us.

No matter how varied the forms of life on earth, whether human, worm, plant, or bacterium, all these living creatures are the result of one and the same biological starting situation, all are related to each other. But it is not unlikely there are forms of life we cannot imagine, forms that have absolutely nothing in common with us and with which we can find absolutely no basis of communication.

This is by no means a plea to break off our attempts at establishing contact with other intelligent beings in the universe. It is merely a reminder that man at the bottom of his heart is still a Ptolemaist: he considers himself the center of things. The quest for other intelligent beings is currently inaccessible to theory. Should we establish contact, by whatever means, with such creatures, the consequences that could result from it would be unpredictable. Be that as it may, the effort of continuing this age-old search is definitely worthwhile.

Is there other life, perhaps other rational beings comparable with men, elsewhere in the universe? This question has occupied the thoughts of man from the time his idea of the universe made it feasible. The thought did not arise when man considered the planets and moon merely lights and the stars golden nails in the vault of heaven. But as soon as at least the moon and planets were accorded a corporeal nature the idea became inevitable.

There are many examples to be found in the history of philosophy that this question exercised the human imagination at an early stage—sometimes affirmatively, sometimes negatively. Thales of Miletus (about 640–550 BC), for instance, spoke of stars in terms of "other worlds"; his pupil Anaximander (about 610–545 BC) was, as far as we know, the first to claim that the number of worlds was infinite and the sun's size was the same as that of the earth. The Greek poet and philosopher Xenophanes of Colophon, a contemporary of Pythagoras who lived during the sixth and fifth centuries BC, thought that the moon was inhabited and that there was an infinite number of worlds, although not at one and the same time. Xenophanes scoffed at the anthropomorphic gods his compatriots worshiped; he believed that cows, if they could make idols, were bound to worship cattle.

Anaximenes of Lampsacus, a friend and companion of Alexander the Great, told his master to his amazement that he had conquered only one of many worlds. Democritus, to whom we owe the first atomic theory, also postulated an infinite number of worlds. And in *De Facie in Orbe Lunae,* Plutarch described the moon as an inhabited world.

Even representatives of Christianity, such as Origen (AD 185–253), propagated the concept of a continuous creation, constantly repeated, of the world, and saw mortals as astral beings after their death on earth. During the Middle Ages such ideas were suppressed by the Church with its established theory of a geocentric world; they did, nevertheless, continue to find expression. Thus Nikolaus von Kues (1401–1464), a respected churchman, presented the view that the universe was infinite and that creatures comparable to man might therefore exist on other celestial bodies. Giordano Bruno (1548–1600) had to pay with his life at the stake for his conviction, among other misdemeanors, of the multiplicity of the worlds and that these are inhabited. The discussion continued, in various forms, in the widely known utopian novels of the eighteenth and nineteenth centuries.

From a natural-scientific point of view it has been possible to approach this question only in our own century. Many events stimulated man's interest in it during the closing decades of the nineteenth and the opening decades of the twentieth century. There was Schiaparelli's discovery of the "Martian canals" in 1877. Those fine lines, straight as a die, which the Milan astronomer thought he could see on Mars through his telescope, intrigued both experts and the public for decades. The first ideas were formulated and speculations arose about space travel and a number of discoveries were made with ever larger and better telescopes. All this naturally gave an enormous impetus to human imagination.

FROM BINOCULARS TO GIANT TELESCOPES

From the beginning of the seventeenth century the history of astronomy is largely a history of instrumental aids. A new sky opened up when Galileo pointed his first primitive telescope at the stars. But this was not so much a revolutionary event but an *evolutionary* one: all the fruits the invention of the telescope yielded in astronomy have not yet been gathered. Although we are able today to send research probes into space and to other celestial bodies, the telescope is not at all obsolete, and we cannot do without it. Especially the last few years, after a period of stagnation, have seen new improved telescopes built and under construction around the world. Space travel is no substitute for earthbound, terrestrial astronomy; it is a useful supplement in the same way that radio astronomy and optical astronomy, far from competing with each other, are methods of research that support and supplement each other.

The flowering of optics and of precision mechanics during the nineteenth century is last, but not least, the result of the astronomers' demands for ever larger telescopes and ever more precise mountings for them. Thus telescope construction became the pacemaker of contemporary technology in a way similar to space travel within the last few decades, which brought with it a tremendous spin-off in electronics and computer technology, in propulsion design and material science as well as in many other disciplines.

The building of new observatories, such as the European observatory at La Silla in the Atacama Desert of Chile and the German-Spanish Astronomical Center of the Max Planck Gesellschaft on Calar Alto in Spain, is the result of astronomical planning and consideration. These observatories are necessary to complete the worldwide observation network and at the same time to help the as-

The dome of the 2.5m (100in) reflecting telescope at Mount Wilson Observatory in California. It was erected in 1917 and for almost thirty years was the world's largest telescope. Many outstanding discoveries, especially of other galactic systems, were made here.

tronomers observe and take photographs of stars in more favorable conditions. Progress in optics, which has produced telescopes based on novel design principles, new materials for telescope mirrors, and so on, often contribute to the desire for new instruments, although even now a fairly large number of instruments built during the second half of the last century are still in use. The "characteristic" life of an astronomical telescope can therefore be rated at more than fifty years. The active life of such ancillary instruments as spectrographs and cameras is shorter, about twenty years. Electronic devices, not least because of the continuous progress achieved in this sector, are as a rule discarded after five years, to be replaced by more up-to-date equipment. Most of the great famous observatories are no longer "modern" according to present standards. For instance, the 5m telescope on Mount Palomar, the largest telescope in the West, was set up in 1948. It therefore was not equipped with the perfect electronic device for the lining up of the objectives, for "tracking" (the telescope "tracks" the apparent motion of the sky) as, for instance, the 3.8m reflecting telescope on Kitt Peak in Arizona. The 200in mirror, as the 5m instrument is also called, is nevertheless still in great demand as a research instrument by astronomers. Even veterans such as the 102cm refractor of the Yerkes Observatory—the world's largest refractor, built in 1897—continue to play a very useful part in observational astronomy.

The quest for new observatories, especially in the Southern hemisphere, is due to the fact that the Southern sky has been explored to a far lesser extent than the Northern; there are, however, a number of highly interesting objects in the Southern sky, such as the Magellanic Clouds, two accumulations of stars in the vicinity of our galactic system. The Magellanic Clouds have many properties typical of large galaxies, but because of their proximity they can be investigated in much greater detail. Today, fundamental exploration of the Southern sky is in progress with the 3.6m reflecting telescope and other instruments, such as special devices for celestial photography. The project of the European Observatory in the Southern hemisphere is supported by the governments of Belgium, the Federal Republic of Germany, Denmark, France, the Netherlands, and Sweden.

Just as busy are the astronomers of the German-Spanish Center of the Heidelberg Max Planck Institute, aided by the favorable weather conditions on Calar Alto. Currently, the largest instrument on Calar Alto is a reflecting telescope whose main mirror has a diameter of 2.2m. An even larger instrument, with a 3.5m mirror, is under construction and expected to be operational in 1983. The focus of this observatory is the investigation of the evolution of the stars—their sequence of birth, life, and death—as well as infrared astronomy—the investigation of the infrared radiation of the stars (if it is at all possible from ground level) with special instruments.

These two observatories have done much to help European astronomy catch up with its American counterpart. In the past practically the only large instruments were found in the United States, and the observatories there had been installed in areas of favorable weather conditions; most European institutions were unable to compete from this aspect. But today, the Europeans, too, are in a position to produce good astronomical results thanks to the siting of the European Observatory in the Southern hemisphere and of the German-Spanish Astronomical Center in areas that are meteorologically and astronomically favorable.

The frequently voiced opinion that new astronomical telescopes are no longer needed at ground level since the introduction of space travel, radio astronomy, and balloon astronomy, misinterprets the facts and is totally wrong. Again, these different branches and methods, far from duplicating, supplement each other.

WINDOWS INTO SPACE

When, during the 1950s, it became necessary to explain to Konrad Adenauer, the first Chancellor of the Federal Republic of Germany, why Germany must have space research with balloons, rockets, and satellites, the task fell to Goettingen geophysicist Julius Bartels. Bartels pointed out to the Chancellor that humans are really like fishes; they keep close to the bottom of the sea, never go beyond the surface of the water, and are therefore denied a realistic view of the whole world. The light of the sun penetrates the depths in which they live only through a filter. Therefore, all they could see down there would be a diffuse brightness from above, not the sun's disk, not the moon, not the stars. Mankind, Bartels explained, is in a similar situation, living at the bottom of an atmosphere that distorts, if not altogether hides, many of the cosmic events: man sees only sections of a very wide range of processes. Adenauer, always open to comparisons that make sense, liked this parable; and Bartels received the funds he wanted for space research.

Man is indeed awkwardly placed for observing events in space. There are electromagnetic vibrations or waves, which establish numerous interactions between the objects of our world. They include not only visible light, but also radio waves, infrared and ultraviolet light, x-rays, and gamma rays. All these electromagnetic waves differ only in their length, or frequency—that is, in the number of vibrations per second. The shortest waves (the gamma rays emitted by radioactive substances and during high-energy reactions in the universe) measure only five trillionths of a centimeter, the longest waves 18,000 kilometers. All the electromagnetic phenomena known to us occur within this huge spectrum. But when we look up at the sky we find that there are only two tiny regions in this spectrum, two windows through which we are able to look into the universe and electromagnetic rays from space reach us. One window is in the region of visible light, the other in that of radio waves of the m and cm band.

Visible light, with its colors red, yellow, green, blue, and violet, can pass through the atmosphere and consists of waves from 8 ten thousandths to 4 ten thousandths of a millimeter. The radio band of waves that can pass runs from 1cm to about 10m length. Everything outside these regions is reflected or absorbed by the atmosphere.

Astronomers were unable, for instance, to observe the sun in shortwave ultraviolet light until it became possible to send suitable observation instruments up to great heights in balloons. The higher the altitude, the wider the window opens in the ultraviolet region. The situation is similar in other regions of the spectrum. But for the investigation of some of them it is necessary to leave the earth's atmosphere altogether. This applies to the x-rays emitted by the sun and the stars. Pictures of the sun in x-ray light are physically of extreme importance because they show us regions of the sun's atmosphere that emit very hot radiation.

Significant regions of the electromagnetic spectrum, which extends from long radiowaves through short waves, heat radiation, and visible light.

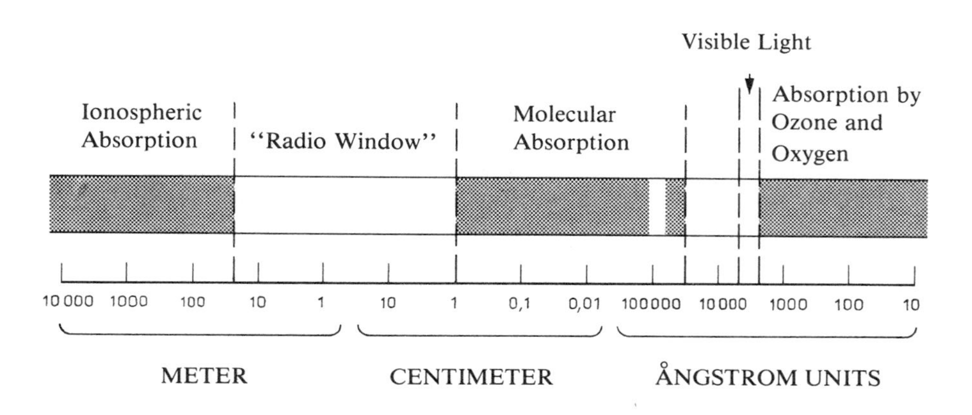

Ultraviolet light, x-rays, and gamma rays are held back by the earth's atmosphere. As you can see, there are only two "windows" into the universe through which we are able to see.

Isolated items of information about x-rays were obtained as far back as the beginning of the 1960s by means of high-altitude research rockets and satellites such as OSO (orbiting solar observatory) and SOLRAD (solar radiation). Some Soviet satellites were also able to pick up these signals. The first photographs of the whole sun in x-ray light were taken on board the U.S. space station Skylab in 1973 and 1975. X-ray astronomy has now become a separate branch of astronomy. It also includes x-ray emission of the stars, which has yielded a number of fundamental findings.

The telescope "Stratoscope II" (shown after its first landing) was launched in December 1971. Astronomers use balloons to send measuring instruments into upper layers of the atmosphere and photograph and measure objects in wavelength ranges that do not reach the surface of the earth.

THE NEW ASTRONOMY

Summing up we can claim that the possibilities and methods of space travel, the enormous strides made in electronics (most of it a spin-off of this very activity), and in radio astronomy, have given classical astronomy unprecedented momentum in almost every field. The marriage between astrophysics and nuclear physics has had similar effects.

A new astronomy has been created, an astronomy that looks at the phenomena in the universe from the most varied perspectives and investigates some of them (at least in our own planetary system) *in situ* "on location." Astronomy has become an experimental science. Sodium and barium clouds, ejected by research rockets in the upper atmosphere, have yielded information about electromagnetic interactions and diffusion of certain gases. Comprehensive experiments with lunar rocks have been conducted in laboratories on the ground. Magnetic fields and particle radiation, in areas both near and distant from the sun, were measured and their interactions investigated. All this is part of the new astronomy.

Let us now examine in detail the picture of the universe as presented by this new astronomy.

The New Cosmic Home

MODERN EXPLORERS: SPACE PROBES

Magellan's reports of his travels, the description of Livingstone's adventures, the dramatic accounts of Scott's polar expeditions and Amundsen's flight to the Pole have lost nothing of their fascination. The television broadcasts about the adventures of brave men who set out to visit faraway, unknown regions, to explore the unexplored, have a powerful public appeal.

The conquest of the earth—a conquest where the mysteries of the planet are overcome—is practically complete. Thanks to his technical facilities there is no point on this globe that man cannot reach. It is now possible to get a total view of the earth, a view more complete than one offered by an airplane. This picture is provided by geostationary weather satellites stationed 36,000km (a distance about three times the planet's diameter) off in space.

These satellites are not stationary in a certain spot in the heavens. They move eastward at a speed of 3km per second (11,000kmph), parallel to the earth's equator and complete a circle around the earth every twenty-four hours, the same time it takes the earth to rotate on its axis. So these satellites keep in step, as it were, with the rotation of the earth and follow in its rotatory motion. This means that to an observer they would always seem to occupy one place in the sky. They *appear* stationary, hence the name geostationary or geosynchronous.

These satellites transmit the pictures we see on our television screens during the weather forecast. In addition, they provide the experts with much meteorological and physical information. They transmit pictures in selected regions of the spectrum, measure temperatures and water vapor content of the earth's atmosphere, determine the swell and temperature of the oceans. Other comparable satellites, called earth resources technology satellites, record the vegetation on the earth's surface, distinguish between deciduous and coniferous forests, between ice-caps and deserts, even between different kinds of cereal and between healthy and diseased plants. They reveal the courses of rivers, provide data about volume of melted snow to be expected in the spring runoffs, and permit estimates of the total water economy of our home planet.

Thanks to modern space technology the earth is not the only body circled by such satellites; some have already been put into orbit around other planets of our solar system. Viking I and Viking II, for example, have been circling Mars since the summer of 1976, transmitting similar pictures and information about the weather situation on Mars. Other space probes were sent to Mercury, Venus, Jupiter, and Saturn. Over the course of many months they covered hundreds of millions of kilometers and sent back much valuable information about these planets.

Such space probes are technologically highly advanced robots; they obtain their data with complex measuring instruments and send it back to earth by means of electromagnetic waves. Their robustness is both admirable and enviable. To begin with they are well wrapped on top of a rocket shot into space from earth at a velocity of 11.5km per second (more than 40,000kmph). This happens within fifteen minutes, so that they are subjected to extreme forces of acceleration. These probes withstand forces of 12g to 15g and beyond, which means that at the moment of rapid acceleration they weigh twelve to fifteen times more than when at rest on the earth's surface. They must be robust, and their electronics are designed both to resist acceleration and other such forces and still function perfectly. During their long space flights they are weightless; the concepts "top" and "bottom" are irrelevant in space. This calls for certain principles of design. For instance, propellants are stored in tanks and blasted through the jets at radio command from the board computer or from mission control to turn the probe into a certain position. In the condition of weightlessness, these propellants no longer behave as they do when subject to gravity. Their properties change, and so they require different design specifications.

The launching rocket provides the boost to start the probe toward its destination. Once free of the rocket, kinetic energy and the gravitational forces of the sun and the planets propel the probe. These forces can be precisely calculated and are determined by a battery of ground-based computers before launching. The paths along which the probes move are not random but predetermined, the result of extensive calculation.

At the side of the probe facing away from the sun the temperature of the material is 200, 260, 270°C below freezing. The side facing the sun is heated to 100, 150°C, or even higher, depending on the probe's distance from the sun. Ingenious thermal materials are used to keep the electronics inside the probe at a temperature between 15 and 20°C, no matter what the temperature of the outside walls.

Air cooling is out of the question because there is no air in space; this vacuum is another difficulty dealt with in design of these flying laboratories. Radiation is another hazard; not only the intense heat but also the ultraviolet radiation of the sun. Only a negligible fraction of this radiation reaches the earth; if the upper layers of the atmosphere did not absorb it, human life on earth would not be possible. There is also solar wind, a stream of charged particles (mainly protons, atomic nuclei of the element hydrogen), similar particles of cosmic radiation, and occasionally micrometeorites with which the probe may collide. All the various magnetic fields must also be considered.

In spite of all these hazards, and a flying time of months or years before they reach their predetermined destination, these probes function perfectly on arrival, transmitting pictures and data. Only very rarely does such a device suffer a premature breakdown.

Space probes are supplied with electricity from solar energy or from nuclide batteries. Those that are not sent too far out into space but remain within the sun's optimum radiation range, that is, those circling the earth or flying to Mars, Venus, or Mercury, as a rule carry extensive solar cell blades. Semiconductor elements, the

---

Pioneer Venus was one of the most successful space probes. These are the five landers of Pioneer-Venus 2. They were launched from earth on August 8, 1978, and separated on December 12, 1978, to fly through various regions of the Venusian atmosphere. The landers transmitted data about density, temperature, and chemical composition.

solar cells, are mounted on these structures, which look like wings or sails. When they are exposed to the light of the sun they produce electric current, which continuously recharges the batteries on board. Probes sent to Jupiter and to even more distant planets usually carry radionuclide batteries, vessels in which a decaying radioactive element generates electric current. This current is required for the operation of the research instruments of the probe; for monitoring its state; for radio communication with the ground; and for the transmission of pictures taken by it, of scientific data, and of a number of technical control data.

Voyager 1

Let us look at one of these probes in greater detail. Voyager 1, a missile launched from Cape Canaveral on August 20, 1977, flew by Jupiter at a distance of 278,000km on March 5, 1979 and reached and flew by Saturn on November 13, 1980. Its shape is difficult to describe. It is dominated by a dish-shaped aerial with a diameter of 3.7m and a boom length of 13m. At the end of the aerial are magnetic-field measuring instruments. Behind the aerial, a decagonal box, always aligned for transmission to and commands from the earth, contains the electronic instruments, computers among them, which can be programmed by radio from earth.

Voyager's research instruments are mounted on a swiveling platform. An infrared spectrometer records temperatures at various levels of the atmosphere of the planets visited and provides information about the composition of the gases of the atmosphere of planets and of moons. With an ultraviolet spectrometer the chemical composition of the atmosphere is determined; it looks especially for hydrogen and helium. A photopolarimeter reveals the nature of the surface of the moons and tells us about aerosols (finely distributed particles of dust and liquids) in the atmosphere. Instruments measuring charged particles record the energy and forward direction of electrons, protons, and other elementary particles in space, as well as the related particles of cosmic radiation. Various magnetometers continuously measure the surrounding magnetic field.

Interesting relations can be derived from the interactions between the charged particles and the magnetic field. A whip antenna with a length of 10m is used for radio-astronomical investigations; it records especially the radio waves emitted by the planets. Two cameras—one with a long focal length (1.5m), the other a wide-angle one—take the breathtaking pictures we have received of Jupiter and Saturn from Voyager. Three radionuclide generators, driven by decaying plutonium oxide, supply 390W of electrical energy; the heat energy liberated during the radioactive decay is converted into electric current. This wattage is sufficient to power the measuring instruments, to drive the controls, and to ensure radio communication with the earth.

Naturally the electromagnetic waves of the radio transmissions that eventually reach the earth are of such low energy that only the largest space receivers are sensitive enough to record them and, via complicated amplifiers, to convert them into useful signals. The energy such an aerial receives from Voyager is so weak it could not power even a penlight. Yet Voyager successfully transmitted more than 19,000 photographs and much valuable scientific and technical data from its flight by Jupiter.

About $880 million (£440 million) has been spent on the Voyager project so far. More than 10,000 workers have been busy putting it into practice. This space missile, weighing about 800kg, is still hurtling through space at twenty times the speed of a cannonball at a distance so enormous as to be almost beyond imagination. If, for instance, we wanted to cover the 1 million kilometers that separate us from Saturn in the most favorable circumstances, say by express train traveling 100kmph day and night without stop, we would be on our way for 1,500 years. The first attempt to dispatch a probe into space dates back to 1958. During the following twenty years, more than 125 space probes were sent to the moon, Mars, Venus, Mercury, Jupiter, and Saturn, of which almost a hundred were successful.

LUNAR AND PLANETARY EXPLORATION BEFORE THE SPACE AGE

These ventures ushered in the New Astronomy of the planetary system. During the first decades of the twentieth century the moon and the planets were neglected by astronomers; by contrast they were the center of interest of observational astronomy during the first 300 years after the invention of the telescope. Their appearance was recorded in many drawings. Interest in them intensified enormously during the last quarter of the nineteenth century. In 1877 Schiaparelli observed fine dark lines on Mars, calling them *canali*, which was translated into many languages as "canals." But Schiaparelli did not intend to interpret the Italian *canale* as "artificial waterway"; he saw in them natural waterways, rivers. However, the wrong interpretation and the resultant exposition kindled the discussion about the "Martian canals and their builders." It resulted in many heated debates. Did highly developed life exist on Mars? Were there intelligent creatures who built mighty waterways on its surface and dealt with the drought on this arid planet?

Mars, however, was not the only planet to attract attention; there was also the planet Jupiter. Many observers drew and photographed its bandlike banks of clouds. Details of the clouds were recorded and from their motion the period of rotation of the planet and other details of its atmosphere deduced. Similar investigations were extended to Saturn, adorned with its famous rings.

But during the twentieth century the interest of specialist astronomers in the planets waned. On the one hand their attention was focused too intensely on the questions of stellar astronomy; and on the other many decades of lunar and planetary astronomy had taught them that their conventional approach would not enable them to make much headway with the exploration of the planets. In spite of the changes observed in the atmosphere of Jupiter and independently of the apparent seasonal changes of details on Mars, for examples, the aspect of the planets in the telescope had become monotonous in the light of fresh scientific discoveries. Photography, too, had proved a disappointment in the exploration of the planets: the visual pictures, recorded in drawings, of the planets turned out to be far more detailed. Photographs were always somewhat blurred because of the atmospheric turbulence during the exposure.

The moon, too, rewarded the observer making drawings of its finest details more richly than his colleague with a camera. Spectrum analysis—used so successfully on the sun—yielded insignificant results with the moon and the planets. Spectrum analysis is based on the splitting up or dispersion of light into a band of the colors by means of a prism. It is well known that white light is composed of the colors of the rainbow: red, orange, yellow, green,

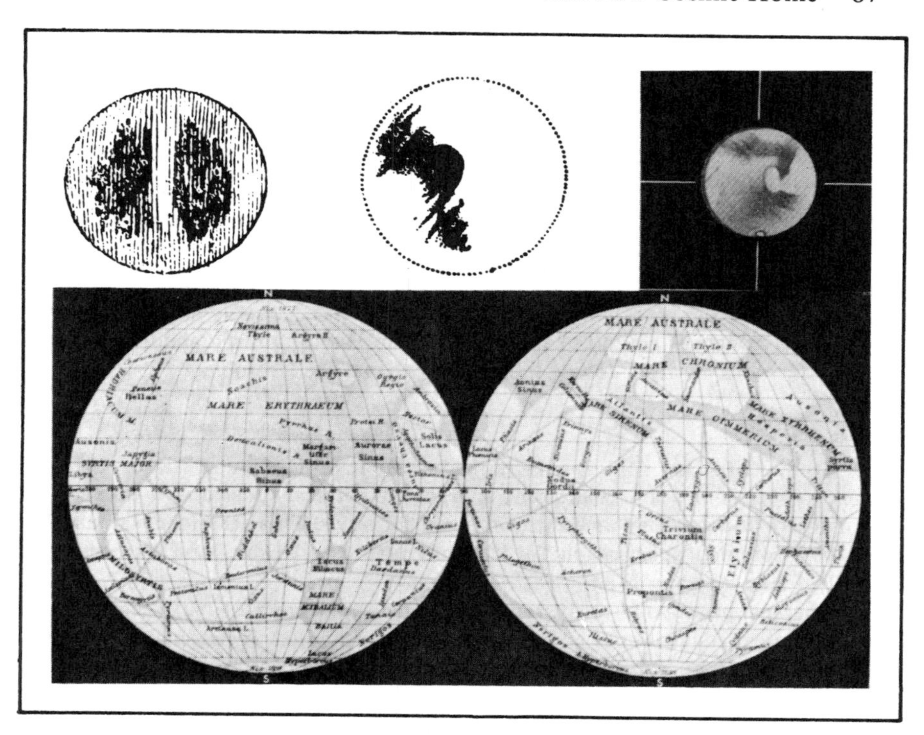

These are examples of observations of the planet Mars. A drawing of Mars by Cassini in 1666 (top left); a depiction by Herschel in 1777 (top center); a modern drawing from 1967 (top right); and Schiaparelli's maps of the Martian "canali" from 1877 and 1888 (bottom).

blue, indigo, and violet. If white light (for instance, the rays of the sun) enters a prism (a glass bar of triangular cross section), it does not pass through it unchanged and reappear on the other side as a patch of white light; it is spread into a band of colors and dispersed, as the expert calls this phenomenon. Sir Isaac Newton noticed this in an experiment performed as early as 1669. If the dispersed colors are made to pass through a prism once again they remain unchanged, are not further dispersed. But if they pass through a collecting lens, they reform into a white patch of light. This proves that the color white is the sum of the colors red, orange, yellow, green, blue, indigo, and violet; these are in fact principal and cannot be further dispersed. But only in 1859 was the great significance of such a spectrum discovered; the physicist Robert Kirchhoff (1824–1887) and the chemist Robert Bunsen (1811–1899) found that it reveals information about the chemical composition of luminous substances.

Every chemical element emits characteristic light of narrowly defined colors and can be identified by these "lines" on the basis of its spectrum, which is, as it were, a kind of "fingerprint" of the element: as a person can be identified by his or her fingerprints, so a chemical element can be identified by its spectrum. Luminous gases produce spectra which consist of individual lines and are therefore called line spectra or emission spectra. But solid or liquid incandescent substances, such as the very dense gases, produce a continuous spectrum, that is, a continuous band of the colors of the rainbow in the form of the solar spectrum.

However, if we examine the solar spectrum more closely we shall find dark lines, as thin as a hair, against the continuous background of the colored spectrum. They occupy the same sites as the spectrum lines of certain substances found in the emission spectra. They are, so to speak, negative images of the emission spectra. These dark lines were discovered by the physicist and optician Joseph von Fraunhofer (1787–1826) and are called Fraunhofer Lines. They were explained by the spectrum analysis introduced by Kirchhoff and Bunsen, which shows that such dark absorption lines are caused by the fact that every substance absorbs precisely those spectrum lines which it emits when it is luminous itself. The dark absorption lines in the solar spectrum indicate the gases present in the solar atmosphere, that is, in those gaseous layers above the light-emitting layer of the sun, the photosphere. The cooler gaseous layers of the solar atmosphere simply absorb that radiation from the lower, hotter layers of the sun and corresponds to the chemical elements also present at higher altitudes.

Spectrum analysis is an extremely reliable, accurate method which detects even minute quantities of a chemical element. But it largely fails with the moon and the planets because, as we know, these bodies do not emit light of their own, but only radiate light "borrowed" from the sun. As a result, the spectrum naturally is a precise image of the solar spectrum; only in the upper layers of the atmosphere of the planets can absorption take place of certain spectrum lines by the gases present there. Even before research based on space probes became possible, spectrum analysis showed that the atmosphere of Jupiter must consist to a large part of methane and ammonia: this fact was eventually confirmed by the analyses of the

Pioneer probes. Spectrum analysis does not work with the moon, because it has no atmosphere.

This was the state of affairs about the moon and the planets before space probes with their new possibilities and methods were introduced. It is hardly surprising that astronomy during the first decades of our century concentrated greater efforts on the stars, where spectrum analysis could achieve such exciting results. In addition, new, more efficient, more powerful telescopes continually expanded the frontiers of the universe. The advance of our knowledge of the internal structure of the atom and of the atomic nucleus also contributed to the interest in stellar astronomy. A linkage of astrophysics and nuclear physics established new research methods and thereby revealed fresh information about the physical and chemical states and changes of state of the stars over millions and billions of years. It became possible to theorize about the energy production, life cycle, and longevity of stars, to make and to verify by observation certain forecasts about the structure of specific stars. New concepts about the structure of our galactic system were linked with observations of the structure of other systems. Is it surprising that in view of all these fascinating developments in stellar astronomy the planets were neglected by the experts, and became for decades the almost exclusive domain of a few serious, committed amateur astronomers? This state of affairs changed only with the beginning of space travel.

THE MOON—KEY TO THE PLANETS

The adventure of space travel to the moon began with the flights in October 1959 and lasted until December 1972—at least for the time being. Basically all those involved and interested in these expeditions believe that the completion of the Apollo Program, with the return of the astronauts of the Apollo 17 flight, was not the end of the lunar program but merely a hiatus.

Even before 1959 American and Soviet unmanned probes had headed for the moon. But in October 1959 something of major importance to all lunar explorers occurred; this is why October 1959 is considered the effective date of the beginning of lunar exploration through space travel. The Soviet lunar probe Lunik 3 orbited the moon and, from a distance of 6,300km, photographed its far side.

These were the first pictures man obtained of the far side of the moon. At that time (the younger readers must already be reminded of this) no man had yet orbited the moon. Our assumptions about its far side were pure speculation.

It is true that we can see a little more than half, 59 percent to be exact, of the lunar surface from earth. This is due to irregularities of the orbital motion of the moon around the earth, a phenomenon called libration. This enables us at times to look a little "behind the right and behind the left ear," at others over the "top of the head and under the chin" of the moon. These libration motions reduce the area permanently invisible to us to 41 percent of the lunar surface. Conclusions from our observations of the area of libration along the rim of the moon and theoretical considerations had at least prepared us for what we expected to, and eventually did, see.

The first Soviet photographs of the far side of the moon did not contain a great amount of detail. But they did reveal that the

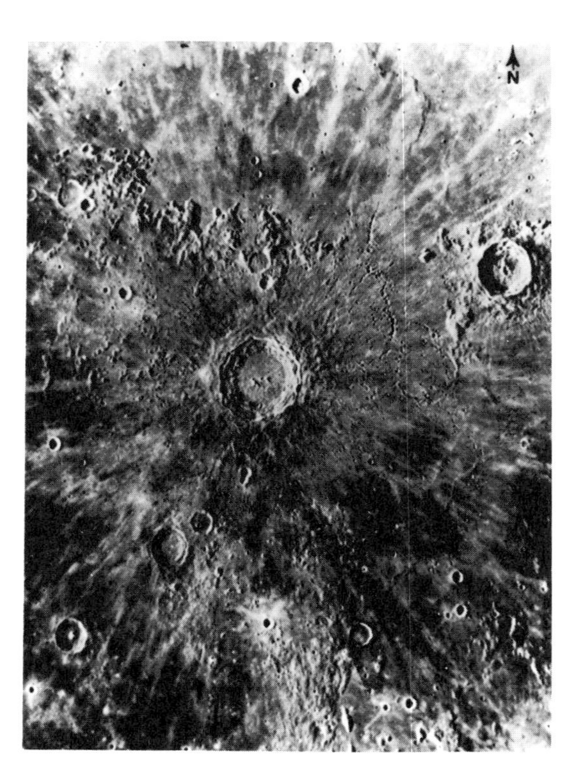

Excellent photographs of the moon were taken from earth during the first half of the twentieth century. They formed the basis for the subsequent mapping of the moon. This photograph shows the ring-shaped Copernicus complex. The crater has a diameter of 90km, and its wall reaches altitudes of up to 3,600m.

moon's far side does not have as extensive lowlands as the near side, but many more craters. This, incidentally, had already been predicted on the basis of theoretical considerations. The picture we now have of the moon has been established mainly by the investigations of space travel. The early Soviet and (to begin with, unsuccessful) American efforts were followed between 1964 and 1966 by the highly successful American lunar probes of the Ranger, Surveyor, and Lunar Orbiter types, which took us a considerable step forward in our knowledge of the moon.

During the decades, indeed centuries of telescopic observation that preceded space travel to the moon, the near side had, needless to say, been comprehensively mapped by means of direct visual observation; a large number of excellent drawn and photographed lunar maps and atlases were available. In the course of this systematic mapping of the moon, thirteen dark-hued lowlands had been found on the near side. The seventeenth century astronomers, on the erroneous assumption that these features were sheets of water, called them *maria* or "seas," a term that has persisted to our day although we have known for quite some time that there is no water on the moon. The moon is indeed one of the most arid bodies in the whole solar system. However, the fanciful names for these plains have survived: the Mare Tranquillitatis (Sea of Tranquillity), the Mare Imbrium (Sea of Rains), the Mare Humorum (Sea of Moisture), the Mare Nectaris (Sea of Nectar), the Mare Fecunditatis (Sea of Fertility), the Mare Crisium (Sea of Crises). The largest of these plains was thought to be an ocean, the Oceanus Procellarum (Ocean of Storms).

The very first observers of the moon (including Galileo) could not fail to see round structures outlined by circular walls. The large ones (the largest, Clavius, has a diameter of 240km) are called walled plains; the smaller ones, ring-shaped mountains; and the smallest, craters. More than 30,000 such objects have been identified from the earth through telescopes; the smallest that can be observed have a diameter of at least 1km. Of course, many more craters have been identified by unmanned lunar probes and by manned space flights. It can be assumed that there are over a million craters with diameters of 1m and more on the moon, and vast numbers of even smaller craters. Microscopically small craters, with diameters of a fraction of a millimeter, have been found on rocks brought back by astronauts. These tiny craters were caused

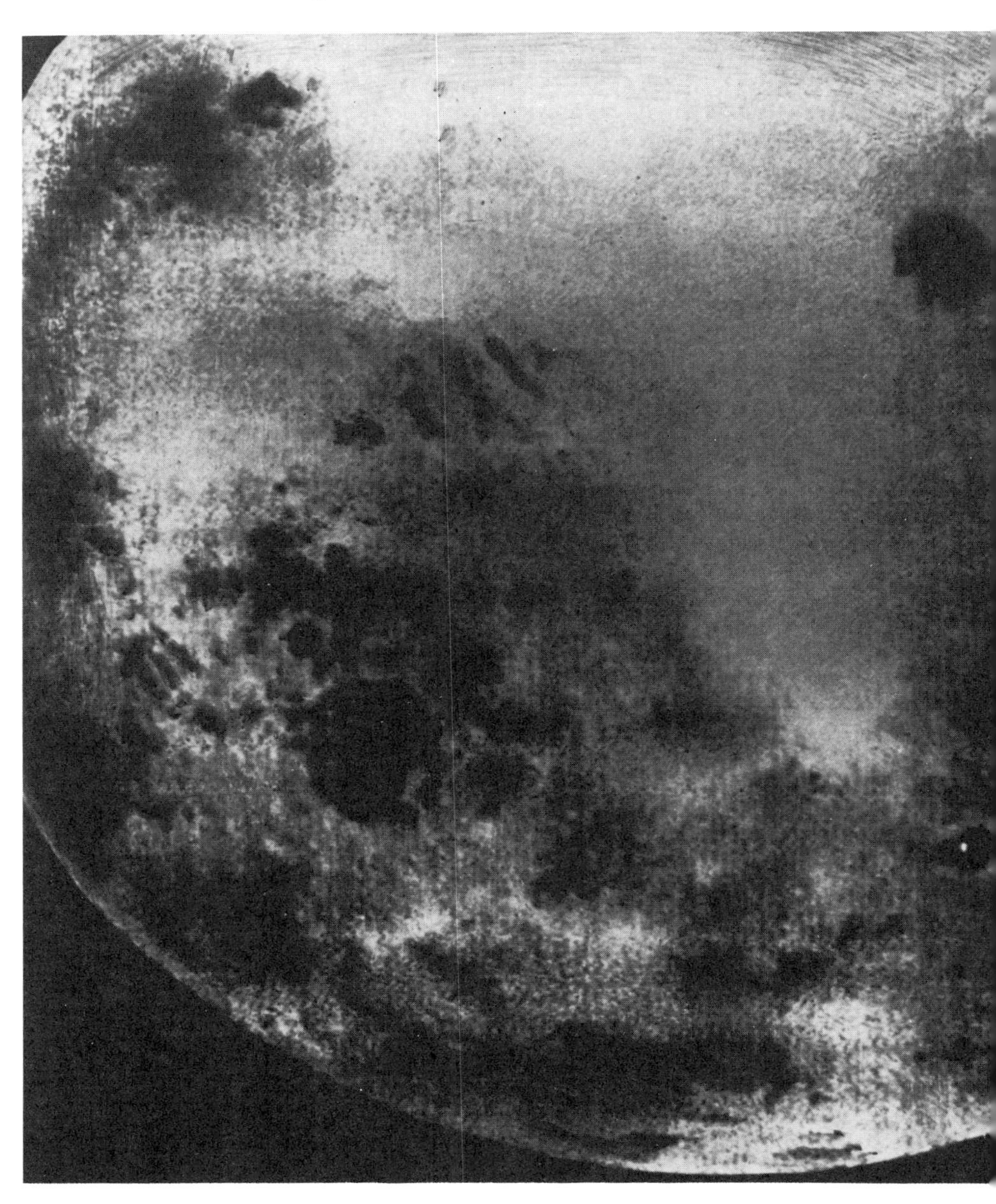

This is the first picture ever taken of the far side of the moon. It was transmitted in October 1959 by the Soviet lunar probe Lunik 3. Although the details lack sharpness, few lowlands are revealed—maria—on the moon's far side.

by micrometeorites, meteoric dust that has bombarded the moon continuously for hundreds of millions of years.

Besides craters, we have found mountains and mountain ranges of dimensions quite comparable with those on earth. Due allowance has been taken for the fact that the moon's diameter is 3,476km opposed to the earth's equatorial diameter of 12,756km. Nevertheless, the Lunar Alps (most lunar mountains are named after terrestrial ones) reach a height of 3,660m; the Caucasus, 5,640m; and the Leibniz Mountains, as much as 9,150m. By comparison, Mt. Everest, the earth's highest mountain, reaches only 8,848m. If we make allowances for the respective diameters of earth and moon, the equivalent on earth of the Leibniz Mountains would be about 31km high and would rise into the earth's stratosphere.

Apart from the tiniest craters, the entire surface of the moon was adequately mapped within the last few years. In fact, today we have better maps of the moon than we have of many regions of the earth. But the cartographic aspect is only one small (but significant) facet of moon study. There are still many questions left open about surface quality, for example, and the evolution of the moon. Space research and travel, especially the landings of astronauts on the moon, have remedied this situation. It is true that even now by no means all the questions about the moon, its importance, and its evolution, have been answered, but knowledge about our satellite has become far more extensive. One of the leading lunar theoreticians of the last twenty years, the Nobel Prizewinner Harold Urey, rashly claimed before the beginning of the Apollo Program: "Give me a rock from the moon and I shall tell you all about the origin of the moon." Some of Urey's opponents sneeringly point out today that we have by now obtained a large quantity of such rocks, but we still do not know how the moon was born.

This statement is incontrovertible, but Harold Urey should not be blamed for his remark. His opinion at the time, voiced on the spur of the moment, could surely be interpreted only metaphorically: lunar rocks investigated in laboratories on earth could provide answers to many vital questions of lunar exploration. There is no doubt about this. It must not be forgotten that we have been practicing geology and geophysics for about two centuries. Man conducted active research on the moon for only a few days when all the visits of the astronauts are added up. The 200-years exploration of the earth's surface and body notwithstanding, geological and geophysical problems are still awaiting solution. Seen in this light the results the moon flights have yielded are of tremendous importance.

Our moon represents an extraordinary phenomenon. Of the approximately forty moons orbiting the planets of our solar system (only Mercury and Venus have no moons), seven can be described as "large." All the others have diameters of no more than 1,000km. With one exception the seven large moons also revolve around large planets, the giant planet Jupiter, whose diameter is 11 times that of the earth, Saturn (9.5 times the earth's diameter), and Neptune (4 times the earth's diameter). In only one case, that of the earth, are the dimensions of the moon and the planet it orbits of roughly the same order. This is the basis of the hypothesis that the

The Soviet lunar probe Lunik 3 provided the first photographs of the far side of the moon. During the probe's return to earth the pictures were automatically developed and transmitted to the ground station by radio.

terrestrial moon is not a true satellite; that it has an independent origin; and that the earth-moon system should be regarded as a double planet rather than a planet and satellite. According to the most recent measurements Pluto and its moon Charon are also a kind of "binary planet."

Moon and earth are, at least at first sight, completely different from each other. The earth has a quite dense atmosphere; the moon has none at all. Water covers 70.8 percent of the earth's surface; the moon, one of the most arid places in the solar system, has no water on its surface. The top layer of the earth, the crust, consists mainly of granite on the continents and of basalt on the floor of the oceans. On the moon we find mainly bright basalts, a high proportion of feldspar, and only a few dark mare-basalts. The earth's crust is only about 30 to 35km thick (although the earth's diameter is almost four times that of the moon), but the moon's crust is 60 to 100km thick, and thicker on the far side than on the near. In addition, it probably consists of two layers, with the boundary between them at a depth of about 20km.

This information is the result of the Apollo Program exploration. Only the measurement of natural and artificially-induced moonquakes made it possible to determine whether the moon has a crust. The greater part of lunar rocks consists of breccia—agglomerates produced by hundreds of millions of years' bombardment of the lunar surface by micrometeorites. Naturally, the earth, too, is subject to such bombardment by small meteorites; however, most of the tiny meteorites burn up or evaporate in the upper regions of the atmosphere. But they do reach the surface of the moon. Chemical analysis of the dust that covers the lunar surface from a depth of a few millimeters to several centimeters has duly shown that about 1 percent of this dust consists of meteoric matter, that is, of micrometeorites that have hit the moon's surface. The soil of the moon, which is being continuously turned over by the bombardment of micrometeorites and by solar wind, is called regolith.

Measurements of moonquakes were carried out for eight years, from 1969, the year of the first landings on the moon, to 1977. The readings were automatically transmitted by the measuring stations set up on the landing sites by the Apollo astronauts and supplied with electricity by a radionuclide generator. NASA, the U.S. space agency, decided to switch these stations off from October 1, 1977,

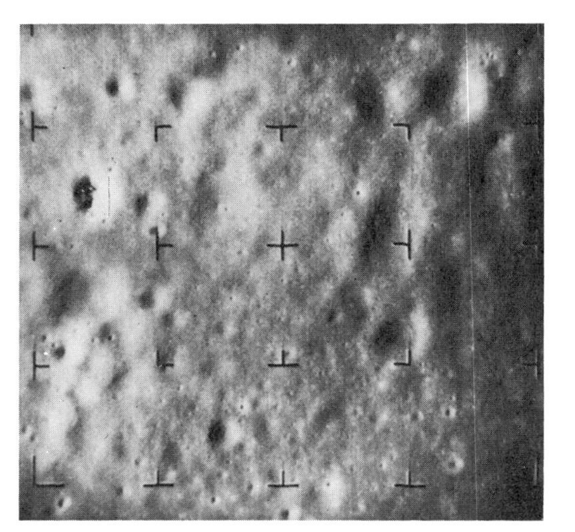

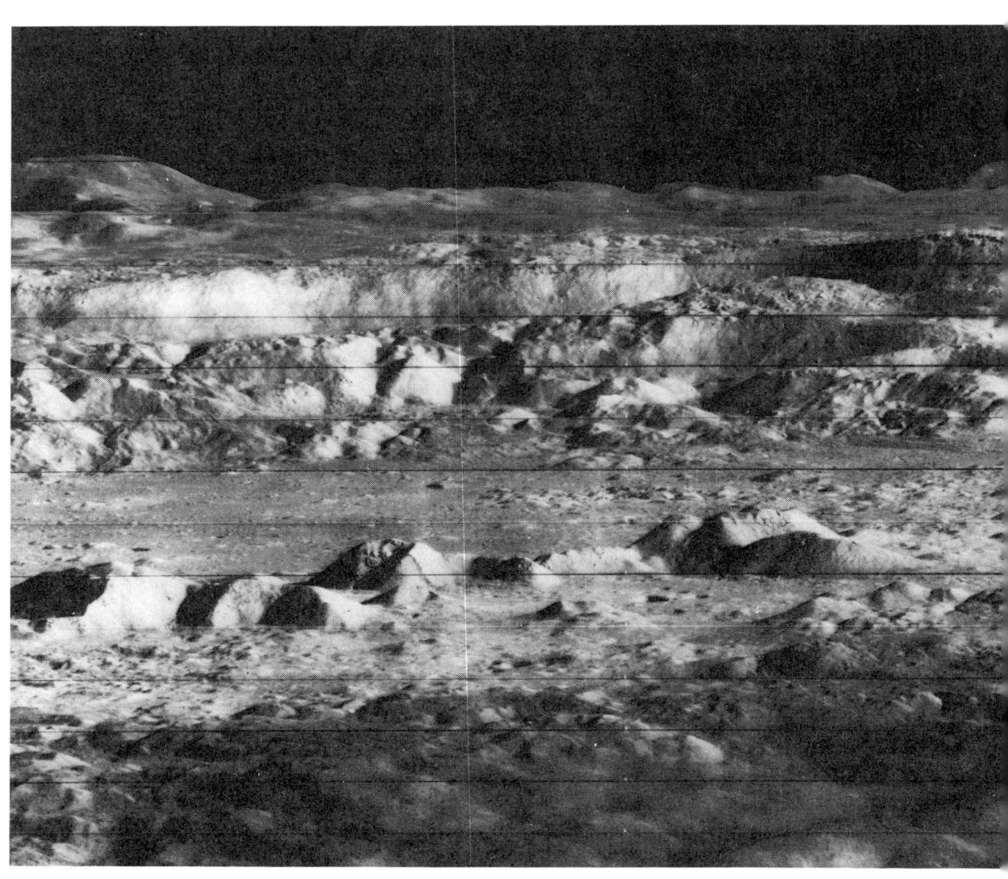

On February 19, 1967, during the search for potential landing sites for manned probes, Lunar 3 photographed this crater on the far side of the moon from an altitude of 1,500km.

Facing page:
Top left: This is one of the 4,308 photographs transmitted by the American space probe Ranger VII before it crashed on the moon. It was taken from an altitude of 48,000m—2.3 seconds before the crash. The smallest recognizable craters have a diameter of about 9m and a depth of 3m. The largest crater has a diameter of 100m and shows a rugged boulder inside.
Top right: Strut of Surveyor 3 after its soft landing on the moon. The automatic grab scooped lunar rock onto the strut where it was photographed closely by television cameras.
Bottom: This photograph was transmitted in November 1968 by Lunar 2 from its orbit around the moon at an altitude of 45km. It shows part of the crater Copernicus. The distance between the features along the front of the picture and those along the rear is about 27km.

because the constant monitoring and evaluation of the technical and scientific data cost NASA about $4 million (£2 million) each year. During operation, however, the stations transmitted more than a billion information units (bits) of data about moonquakes, temperature conditions, heat flux from the lunar interior, electrically charged particles, meteoric impact, and the magnetic field. These data appeared adequate for all investigations and years of further transmissions would no longer have justified the high operating costs. During this period about 10,000 moonquakes and 2,000 meteorite impacts were recorded.

The moonquake measurements have also provided us with the information that the moon has a mantle and perhaps even a core below its crust. The upper mantle extends to a depth of about 500km and has a rather homogeneous composition, consisting probably mostly of olivine and olivine-pyroxine, two minerals also very common on earth. But the composition of the lunar mantle has not yet been fully determined. The shock waves of the moonquakes indicate an enrichment of iron at a depth of 500km. It is, however, still not clear whether the moon has a molten metallic core. A soli-

The lunar crater Tycho, photographed on August 15, 1967, by Lunar 5 from an altitude of 217km. Tycho is an impact crater. Its diameter is 90km, and its walls rise to peaks of up to 5,250m. The central mountain is almost 2,000m high.

tary measuring point indicates the existence of such a core, which if it does exist, should have a radius between 170 and 360km and constitute no more than 2 percent of the moon's total mass. Theoretical considerations also favor the existence of a lunar core. The 10,000 recorded moonquakes within eight years may appear high to those not familiar with seismology, the science of earthquakes. In fact, the moon's seismological activity—both regarding the number and intensity of moonquakes—is extremely weak compared with that of the earth. For roughly 3 billion years the moon has been an extremely inert body, a lifeless world. More than 99 percent of these moonquakes are weak and are triggered by the tidal effects of the earth.They occur at depths between 700 and 1100km at certain periods. Only a few moonquakes per year originate within the topmost 100km of the moon; and their cause is tectonic, the result of movements in the interior of the moon. Unfortunately we still know very little about them.

Aside from their composition, scientists have attempted to determine the age of the lunar rocks brought back by astronauts. Age was measured by the decay of radioactive substances contained in the rocks. This has provided us with direct information about the age of individual rocks, and therefore fix points in the evolution of the moon. These investigations show that the moon, like the earth and the other planets, came into being 4.6 billion years ago. We have already mentioned that there are several hypotheses about how this happened. The so-called *accumulation hypothesis* must be regarded as the likeliest. It claims that the sun, followed by the planets and their moons, formed from a cloud of gas and dust that filled the universe. This process lasted many millions of years and included numerous physical events. The planetesimals, bodies the size of dust particles, occupied the space around the sun and stuck together when they collided; they developed into protoplanets and finally into planets. This hypothesis can be understood only in the context of the evolution of the sun itself, of the prevailing temperatures, of the constrained distribution of the chemical elements, and of some other factors.

Another theory claims that earth and moon formed as a single body and that the moon separated from the earth at a later stage. Some proponents of this theory go so far as to define the area where the moon originated—in the basin of the Pacific Ocean. According to a third, but highly unlikely hypothesis, the moon is a body which came into being in another region of the solar system (or even outside it) and which was captured by the earth.

Whichever idea we accept, the origin of the moon must have been a violent event taking only a few thousand years—in geological terms a mere instant. Furthermore, if we proceed from the thesis that the moon originated in an orbit around the earth filled by millions of planetesimals, dust particles, and clouds of gas, the final stage of the birth of the moon must have been a spectacular event. Huge boulders circling the earth must have crashed into the protomoon and turned the whole structure into a hot, incandescent, molten, continuously growing mass of lava. Eventually the huge boulders accumulated and the moon began to cool.

During the following 200 million years the various materials separated out in the liquid interior of the moon. The heavy elements settled and formed the core; mantle and crust evolved during the same process. Even after the initial continuous bombardment, huge meteorite or planetoid bodies, enormous lumps of rock, struck the cooling moon during the next 500 million years. These collisions left deep scars in the crust and produced the troughs of the present maria.

All this happened mainly on the near side of the moon. The earth's gravitational pull prevented an uncontrolled autorotation of the moon on its axis. The only rotation of the moon was, as now, a synchronous rotation. Even when the moon's crust formed, the earth made its influence felt. The crust on the side facing the earth is thinner than on the far side. As a result, the scars left by the huge boulders that hit the moon on the near side were larger, because the boulders partly pierced the crust. Liquid magma erupted from the mantle of the moon, filling the basins with a layer of lava. These lava basalts today cover about 20 percent of the near side of the moon, but they are only 1 to 2km thick and thus form less than 1 percent of the total volume of the moon's crust. Because they originated at depths between 200 and 500km, they yield interesting information about the interior of the moon, for they are samples, as it were, of the mantle stratum of our satellite.

If we accept the suggestions of American professor Gerald Wasserburg and his group of selenologists, the beginning of mare formation was cataclysmic. Most rocks of the lunar highlands have radiometric ages from 3.8 to 4.5 billion years. This has led Wasserburg to assume that a cataclysmic period of a few hundred million

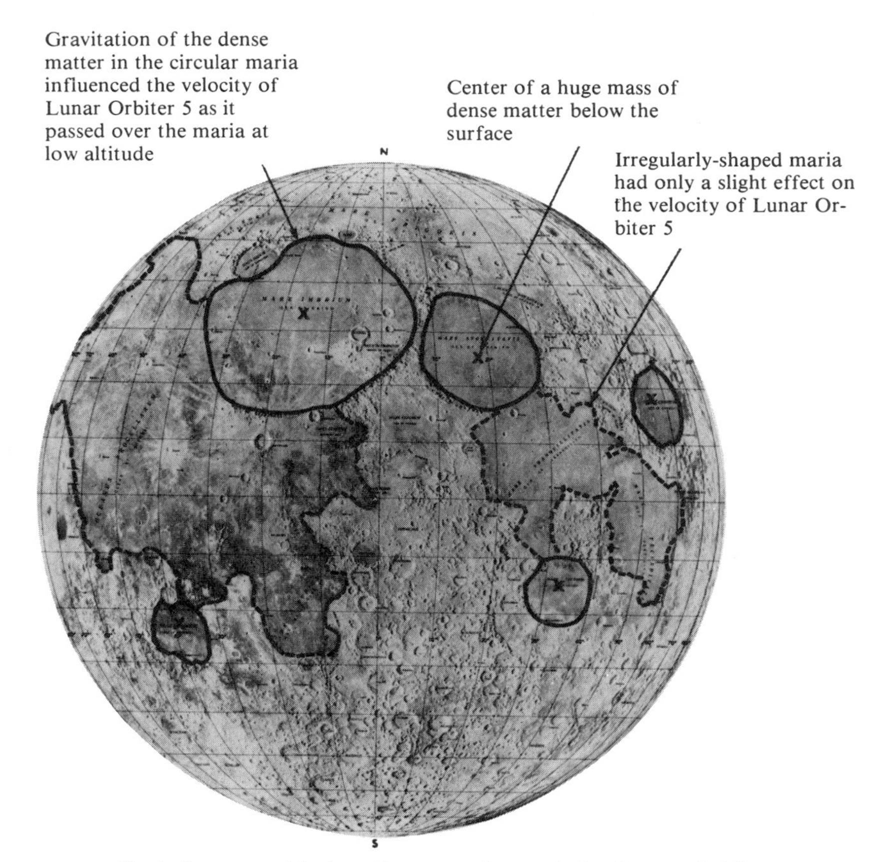

Mascons were first discovered below the moon's maria by Lunar 1. They revealed themselves through an increased gravitational force that perturbed the probe's orbit.

years occurred during the early history of the moon. A large part of the surface rock was melted by the meteorite bombardment and subsequently began to solidify; the radioactive "clocks" of the rocks then began to "tick." In other words, this is the point when the rocks started to mix, and thus the point from which the current proportions of radioactive decay products are counted. Few rocks are left of the preceding period, they are the ones which are up to 4.5 billion years old.

The process of the magma eruption from the interior of the moon into the basins of the maria, however, continued until about 3 or

The Apollo astronauts took many soil samples similar to this drill core (to a depth of more than 26cm). The samples yielded valuable information about the structure of the regolith, the topmost stratum of lunar soil. To date, only about 10 percent of the 385kg of lunar rock collected has been analyzed. The remainder has been set aside and waits for new methods of investigation.

2.5 billion years ago, when the moon ceased to be active. Obviously, further changes took place on its surface; individual meteorites plunged into it, forming as they did the "recent" craters, such as the ray crater Copernicus, formed only about 800 million years ago. Apart from these isolated events, however, there has been deadly silence on the moon for the last 2 to 3 billion years. In contrast with the earth, which is still active, the moon can be regarded as a cosmic museum. Thus, selenology tells us so much about the history of our planetary system.

The oldest rocks found on earth so far are little over 3 billion years old, and even they are rare exceptions. Erosion, wind, and weather plough up the earth's surface; heat and cold, water and ice over the course of millions of years change conditions so that fragments from the early history of our planet can be found only on rare occasions. There is no wind, no rain, no snow to alter conditions on the moon. The moon has no atmosphere to produce weather. At best small quantities of gases, such as krypton, helium, or argon, may be released by the occasional crust moonquakes, but these are local events where cracks may occur in the rocks, or gas evolution take place.

The traces of moon dust reveal information about the composition of the solar wind millions of years ago. Records are found in the high mountains and maria of events that occurred equally long ago. We have so far explored only some of these mysteries. We know that the moon has no general magnetic field. Nevertheless, weak, local magnetic fields have been detected on the lunar surface. This means that the moon must have had a magnetic field of its own in the past or been under the influence of a strong extraneous magnetic field. The various Apollo spacecraft measured the proportions of certain chemical elements, radioactivity, and the character of the terrain. These measurements ascertain that the data obtained at the measuring stations are, on the whole, valid and do not represent random local deviations that would not be applicable to the whole of the moon. Analyses of the 1966–68 orbits of the unmanned Lunar Orbiter satellites have shown that the gravitational pull in the lunar orbit is subject to fluctuations; this phenomenon was confirmed by the Apollo orbits. Mass concentrations—mascons—exist in certain areas of the moon's surface, marked by a greater density of the underlying material. These mascons are interesting indications of the distribution of mass in the interior and of the structure of the moon. They also allow certain conclusions about the moon's early history.

The Apollo Program has yielded an enormous amount of information, of which we were able to outline only a few items here. We owe further data to Soviet space probes, two of which carried a scoop which returned about 100g each of lunar soil samples to earth. Other unmanned Soviet probes had automatic vehicles on board that probed the soil of the moon in the farther vicinity of their landing sites.

All this, however, is merely a beginning. Today the material exploitation of the moon is being seriously considered, such as the mining of its minerals, or the supply of building materials for the construction of a space station orbiting the earth. Such a method would constitute a considerable saving of energy because of the relatively weak gravitational force of the moon. But even for the realization of such plans we know too little about the moon. The appetite of the scientists for further information was really whetted only by the results of the Apollo Program.

New Exploration

Two concrete projects are under discussion concerning the future direct exploration of the moon. One is a lunar satellite to circle the poles of the moon. This Polar Orbiting Lunar Observatory (POLO)

Apollo 15 in lunar orbit. While astronauts David Scott and James Irwin descended to the moon's surface, Alfred Worden continued the orbit and operated several research instruments housed in the space capsule. A synoptic picture of a large part of the moon was obtained through this investigation.

is intended to transmit pictures and physical and chemical data from regions of the moon not yet reached by spacecraft. The Apollo Program, and therefore also the preparatory programs of unmanned lunar probes, were orientated toward the equatorial belt of the moon. The highest latitude reached by a manned spacecraft was 26° by Apollo 15 in the Hadley-Appennine Range in July 1971. The project of the lunar satellite in polar orbit is being keenly pursued by some scientists in America and Europe. Their aim is to launch it as a joint American-European venture as a project of the European Space Agency (ESA).

The second project is of German origin and much more ambitious. The author of this idea is an American astronomer who specializes in the moon and planets and who lives and works in Germany at present. This moon project, called Selene, involves the landing of sixteen unmanned probes in selected areas on both the near and the far side of the moon and the transmission of information from there within the next few years. Eighteen further locations have been selected from which landers/returners are to bring soil samples back to earth. Project author Alan Binder advances several cogent arguments in favor of his program, from detailed scientific reasoning to the indication of material exploitation of the moon. He states exploitation will take place during the next century, but must be based on comprehensive exploratory work.

It is not known if either program will be realized during the next twenty years, of if another, as yet unknown, will succeed. Nevertheless, it is a foregone conclusion that man will return to the moon come what may, first by sending unmanned research and exploration probes, and eventually by sending astronauts. The great adventure of exploring another heavenly body on the spot is not yet at an end. The Apollo Program was just the beginning.

SPACE PROBES TO OTHER PLANETS

The planetary system, its origin, its development, and its present state can be understood only when looked at as a whole. The planets must be regarded as the members of a family, in which each has its own distinct personality and its own specific features. The chemical composition of the planets—the appearance or nonappearance of atmosphere—is the fundamental difference between the Inner Planets (Mercury, Venus, Earth, and Mars) and the Outer Planets (Jupiter, Saturn, Uranus, Neptune, and Pluto). Both time and place of origin were equally important in determining this difference. Temperature at the beginning and during advanced stages of planet formation was a factor that contributed to the end result. The story of the origin and the end product of planet development is closely linked to the evolution of the sun, which took place during the same period, while the protoplanets condensed from interplanetary dust and gas.

The comparison between planets is the key to the understanding of the solar system. This is why our knowledge of the moon is so important. We would be unable to understand a number of results transmitted by probes to Mars and Mercury, for example, were it not for data on the evolutionary history of the moon.

The First Probes

The dispatch of early probes to other planets was a laborious undertaking, often dogged by failure. Let us take the first probes to Mars as an example.

In 1960, the Soviet Union made the first attempts to send probes to Mars. The first two launches by the Soviets, in October 1960, ended in failure. After a third launch, in October 1962, Sputnik 22—which was to serve as launching platform for the probe unit—exploded in orbit around the earth. The fourth attempt, on November 1, 1962, was the launching from Sputnik 23 of the Mars 1 probe. Mars 1 left its orbit around the earth and entered the flight path to Mars, but on March 21, 1963, radio contact with the probe was lost. The fifth attempt also ended in explosion.

The first and second American attempts to fly by Mars were made in November 1964. The Mariner 3 probe did fly by the red planet, but the protective cap had not detached itself from the payload, no radio contact was established, and the project had to be regarded as a failure. Mariner 4, at long last, achieved the first success: the probe flew by Mars at a distance of just under 10,000km and transmitted physical data and twenty-two pictures of the Martian surface.

Thereafter the Mars program became consolidated. Naturally, failures still occurred, but the tasks became more ambitious. The aim was no longer simply to let probes fly by Mars at a close distance; efforts were made to put them into orbit around Mars and thus make them artificial "moons." The first such success was achieved by the Russians with the Mars 2 missile in November 1971. Unfortunately, a Mars lander it had on board crashed into the surface of Mars during the landing approach. The climax of the Mars program so far were the landings of the two American Viking probes on Mars in July and September of 1976. But before turning to a discussion of Viking let us spend a little time on probes sent to other planets.

The history of the exploration by probes of the earth's second neighbor in space, Venus, is similar to that of the exploration of Mars. It was the Russians who made the first attempts to reach Venus, beginning in February 1961, launching two probes within eight days. The first did not leave orbit around earth; the second did fly by Venus at a distance of 100,000km, but radio contact with the earth had been lost by then. After further Soviet and U.S. failures the first success came with America's Mariner 2 in December 1962: the probe passed Venus at a distance of 34,830km and sent back data about the Venusian atmosphere.

It would be beyond the scope of this book to describe each effort made in the following years. Suffice it to say that from 1969 most attempts were successful. Soviet probes managed to land on Venus despite its hostile atmosphere, whose ground density is more than ninety times that at ground level on earth. In February 1974 Mariner 10 flew by Venus at a distance of only 5,760km and transmitted 6,800 pictures that clearly revealed the structure of the Venusian cloud cover. In December 1978 the Americans at last succeeded in establishing a Venus satellite. Five landers descended from the lander to the planet's surface and transmitted information about temperatures, pressures, and chemical composition of the levels atmosphere to the ground.

From the technological standpoint the development of the space probes can be divided into two stages. The missiles that undertook flybys of Mars and Venus during the 1960s and the beginning of the 1970s represented the first generation of space probes. These flights provided enough experience to go a step further and to design a second generation of space probes that were extremely sophisticated technologically. In the meantime sufficient mastery of rocket flight technology, electronics, and orbital mechanics had been acquired to reduce the expected failure rate. Complicated, costly flight projects therefore became feasible without undue risk of fundamental error in carrier rocket design or failure of a relay that would turn the probe into a swirling ball of fire only a few minutes after liftoff. Although it cannot be entirely ruled out even now, the likelihood of such a mishap is so remote as to lower the risk to a tolerable level, one acceptable even to insurance companies.

The space probes of the second generation are the Viking Mars probes, the Pioneer Jupiter probes, and Voyager, which passed Saturn in November 1980. The two earlier Voyager probes had obtained and transmitted breathtaking color pictures and valuable physical information about Jupiter.

THE NEW IMAGE OF MARS

Mars is a typical example of how advancing techniques of exploration and fresh knowledge can cause prevailing opinions to change not once, but several times. Our ideas of this planet have had to be adjusted to allow for almost every new astronomical discovery concerning it from the end of the 1940s to the middle of the 1960s. The enthusiastic speculations about rational beings on Mars became commonplace with the discovery of the canals by Schiaparelli in 1877 and persisted for decades. As we have seen, Schiaparelli did not consider them to be artificial waterways, but this meaning was generally attached to them because of mistranslation. The clever Martians, ran the explanation, built these artificial waterways to conduct the melted snow or ice from the polar regions of Mars to the equatorial belt to overcome the lack of water in the temperate and tropical zones.

One fervent protagonist and follower of this hypothesis was the American millionaire Percival Lowell. In 1894 he had a private observatory built at Flagstaff, Arizona, where he devoted himself to the study of Mars and its canals. Together with a few coworkers, he made thousands of drawings of the red planet. By 1900 no fewer than 400 canals had been counted. Lowell had begun his exploration with a 24in. refracting telescope (that is, the diameter of the objective was 24in. or 61cm) and continued with a 42in. reflecting telescope, which had as the image-forming component a 107cm concave mirror with a silvered surface. But now observers with smaller telescopes also began to see and draw the Martian canals which came to assume truly heroic proportions. Under Percival Lowell's leadership several groups of astronomers championed the reality of these canals; other groups denied their existence with equal force. The great controversy over the Martian canals had begun. It found an enormous response among the public.

Many experts disqualified the Martian canals as optical illusions. The British scientists Maunder and Evans showed pictures of Mars—on which oceans, continents, and other details had been entered—to school children, and asked them to reproduce the pictures from memory. When the children represented in their drawings the fine details by straight, sharp lines, the two scientists considered it proof that the Martian canals in reality represented hundreds of fine details that the eye was unable to resolve. We know today that this explanation is correct. The eye tends to interpret fine details as geometric patterns, particularly when the details seem to change shape and intensity as a result of atmospheric disturbances.

During the 1920s measurements in the telescope with ultrasensitive thermometers, thermocouples, provided us with the first information about temperatures on the surface of Mars. They proved lower than those on earth, yet were still within the life-sustaining range. Estimates were between 25°C on the equator and −135°C in the polar regions.

Spectrum analysis of the atmosphere of Mars also produced contradictory results. Slipher, in 1908, thought he had found absorption bands of oxygen and hydrogen in the spectrum. Investigations by other workers during the following decades yielded inconsistent information: sometimes oxygen was detected, sometimes it was not. In 1947 Gerald Kuiper at the McDonald Observatory in Texas found traces of carbon dioxide in the Martian atmosphere, and in 1948 he discovered water ice in the spectrum of the polar caps. In 1956 and 1958 William Sinton found absorption lines in the spectrum of the dark regions of the surface of Mars. He interpreted this as an indication of the presence of carbon; from this, in turn, he inferred the existence of organic compounds on Mars. But disillusionment about this latest "discovery" came in 1965 when the two spectrum lines were found to represent a special compound of heavy water, hydrogen deuterium oxide—HOD. Seasonal changes in color were observed in the dark regions of Mars and, naturally, were first explained as showing the presence of vegetation. They can, however, also be interpreted in terms of inorganic changes or variations of the mean particle size and therefore do not indicate vegetation on Mars.

In 1965 the first pictures of Mars were transmitted to earth by Mariner 4. They showed just under 1 percent of the planet's surface with several well-preserved craters and no evidence of erosion. This supported the picture of a biologically inactive planet by showing it to be geologically inert, too.

In contrast, we received during July and August 1969 a total of 202 pictures of the Martian surface from Mariner 6 and Mariner 7. These pictures showed not only heavily eroded craters but also absolutely chaotic regions of ridges and valleys. This aspect was confirmed by Mariner 9 at the beginning of 1972. The more than 7,000

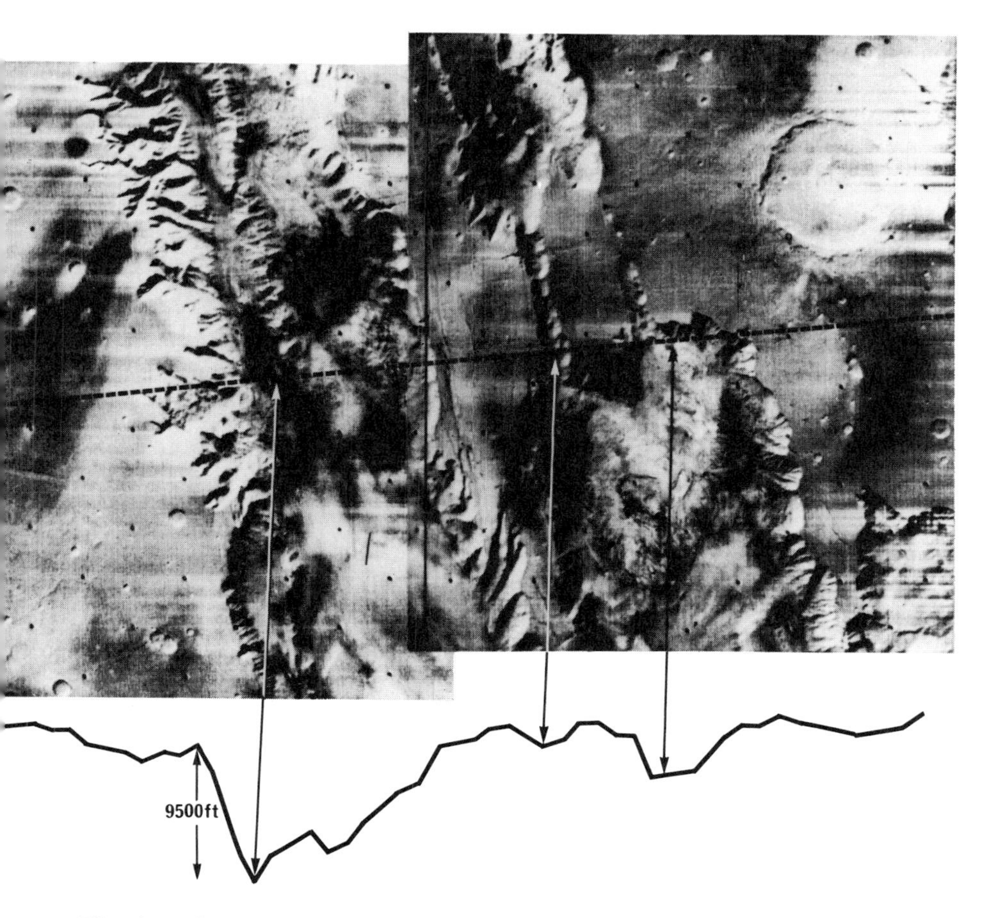

The American space probe Mariner 4 discovered craters on Mars in July 1965. This photograph was taken from a distance of about 9,500km. The discovery of Martian craters led to numerous speculations about the origin of both these and lunar craters. Another probe has revealed that there are also craters on Mercury.

Facing page: This photograph was taken on February 24, 1972, when Mariner 9 was orbiting Mars at a distance of 1,700km. It shows a narrow valley, twice as deep as the Grand Canyon. The elevation profile was obtained through radar scanning and indicates a vertical difference of 2,850m at the lowest point.

pictures of Mars transmitted by Mariner 9 show dry river beds, valleys, ridges, and huge volcanoes. This finding was verified by the two Viking probes.

Each of the Viking probes consisted of two sections. After entering orbit round Mars as a single unit, the lander detached itself from the orbiting section. The orbiters, artificial satellites of Mars, were, like the landers, equipped with their own cameras and measuring instruments. The Viking landers not only gave us a detailed image of the structure of the Martian surface, they also measured the temperatures at all times of the day and night, determined the density of the Martian atmosphere, and analyzed its chemical composition. Also, they used three different methods to search for traces of life. The Viking probes operated from two landing sites 7,300km apart; one 22.27° North, in the planet's tropical zone; the other 47.89° North, in its polar region.

A wealth of information was provided by some 30,000 pictures taken by the orbiters from distances between several thousand and a few hundred kilometers from the surface of Mars. Based on the evidence provided by these shots, the pictures obtained by the landers can be interpreted on a large geological scale, and thus we can form a more complete image of Mars. The most important facts we have gleaned from these orbiter pictures concern the evolution of Mars. There can be no more doubt that water must have been present in huge quantities at an early stage on Mars, with the obvious implication that the density of the Martian atmosphere must at one time have been considerably greater than it is now. At present, gas pressure on the surface of Mars is barely 8 millibar, not even 1 percent of the atmospheric pressure of earth. Under such low pressure water in the liquid state can exist on Mars only temporarily, in minute quantities, in very low-lying areas—if it can exist at all. But the orbiter pictures reveal huge dry river beds and a terrain largely shaped by water, such as ridges traversed by river beds where, in the remote past, water must have forced its way through.

In the summer of 1976 the Viking probes discovered several volcanoes—of dimensions considerably larger than terrestrial volcanoes—on Mars. Although extinct, they yielded interesting information about the evolution of this "unearthly" planet.

Ancient lava flows have also been identified in many pictures. Volcanic activity on Mars must at one time have been enormous, as evidenced by the remains of colossal extinct volcanoes. By comparison, all their counterparts on earth are small. The highest volcano on Mars, Olympus Mons, is 25km high with a base diameter of 600km. It dwarfs Mauna Loa, the earth's largest volcano (4,170m high, with a base diameter of 250).

The North polar regions of Mars were explored and photographed in great detail by the orbiters. The pictures show, for instance, the North polar cap during the summer when it had melted down to a diameter of about 1,000km. Opinions whether the white deposits on the polar caps consist of water ice or of carbon dioxide (dry) ice differed for a long time. Temperature measurements on board an orbiter, however, clearly established that at least the

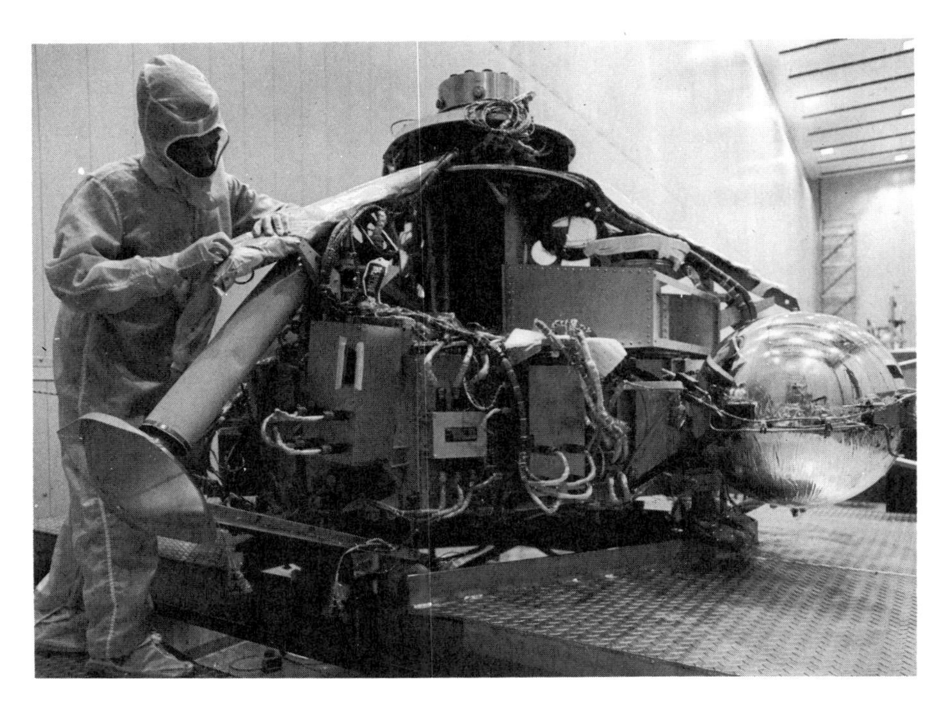

A Viking lander being assembled in the production plant. Because the instruments were to look for extraterrestrial life on Mars they had to be sterilized before launch. The technologists who built the instrument therefore had to wear protective clothing in the "clean" assembly room.

These two pictures of Mars were transmitted by Viking on October 8, 1976. They illustrate methods used in the search for life on this planet. *Left:* the scoop pushes a piece of rock to one side to obtain Martian sand from the hitherto covered spot. *Right:* after the successful attempt the scoop is withdrawn.

residual ice cap consists almost exclusively of frozen water. But during winter, as the cap expands, large quantities of carbon dioxide, now dissolved as gas in the atmosphere, may be deposited on it. This carbon dioxide will evaporate over the following summer. Water vapor is also dissolved in the thin Martian atmosphere, although its quantity is extremely minute compared with that on earth. The atmosphere of Mars is far drier than the earth's. At ground level it consists of 95 percent carbon dioxide, 2.7 percent molar nitrogen, 1.6 percent argon (a noble gas), and 0.15 percent molar oxygen. The average temperatures at ground level measured by the Viking probes range between $-55°$ and $-57°C$. The coldest value recorded is $-85°C$, the warmest $-29°C$. But the Viking orbiter measured $-139°C$ above the South Pole of Mars.

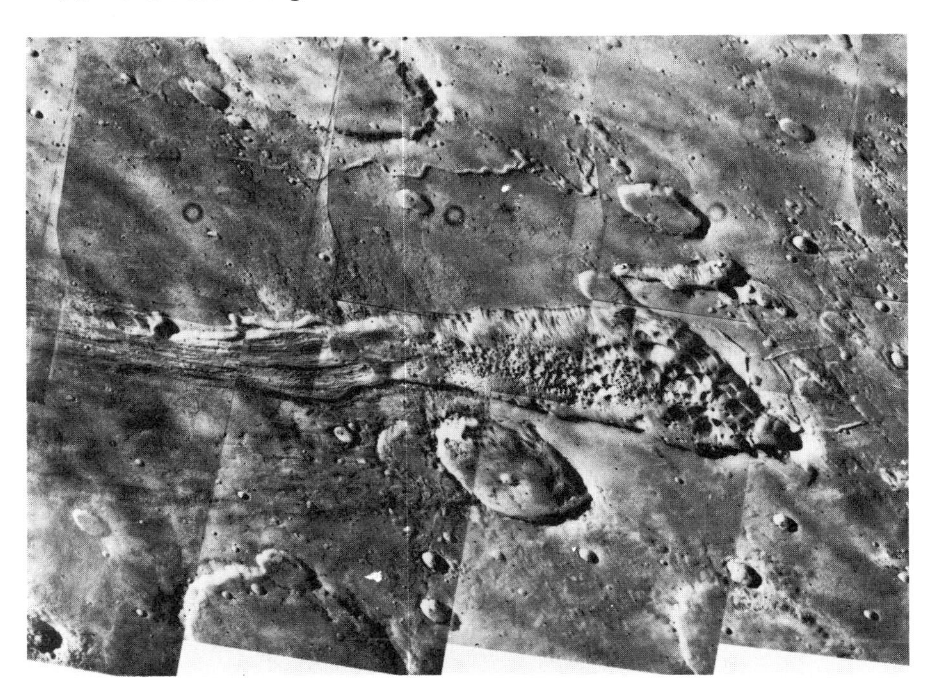

This picture of Mars was taken by Viking 2 from a distance of 2,300km and covers an area of about 300km square. The rugged terrain shows that enormous forces of lava or water must once have been active here.

There is wind on Mars, as shown by occasional violent sandstorms that can last for months. Under normal conditions, the wind, reinforced by gusts, reaches a speed of 9 to 15m/sec at ground level, which corresponds to 32 to 54kmph. During storms, of course, very much higher speeds are reached.

The Viking landers also chemically analyzed the Martian soil. A scoop picked up dust and small pebbles next to the landing site. The automatic analyzers found that matter consisted of 15 to 30 percent silicon, 12 to 16 percent iron, 3 to 8 percent calcium, 2 to 7 percent aluminum, and 0.25 to 1.5 percent titanium.

The results of the search for life on Mars represented one of the climaxes of the project. For some time the results were encouraging indicators of life processes. Today, however, probably everybody concerned has come to accept the view that there is at present no life on Mars. Whether this has been different in the past or will be different in the future is still open to question.

It is quite possible that a chemically extremely active Martian surface totally sterilized the planet and prevented the creation and spread of life. It is equally possible that we used the wrong methods and looked for life in the wrong regions of Mars. Perhaps there is a form of life on Mars that exists only at the poles or only in specific lowlands of the equatorial belt. Perhaps there are other life forms, unknown to us, that do not produce changes in their environment. Proceeding from the thesis that life makes itself known through metabolism—that is, through breathing or photosynthesis—and

Mosaic pictures such as this had never been seen before they were transmitted by Viking 2 on October 4, 1976. They show the North polar region of Mars. In the rugged mountainous terrain about 350km south of the North Pole snow- and ice-covered regions alternate with sunlit highlands free from any snow or ice.

that these reactions somehow affect the environment by absorbing or secreting chemical substances, the instruments on board the Viking probes looked only for a form of life that introduced such changes.

Such a reaction was indeed observed. One of the three experiments to detect life on Mars—the gas exchange experiment—had the objective of detecting respiration processes. For this purpose soil samples, together with a control gas, were introduced into a gastight container, and a nutrient broth was added. The mixture was then investigated for chemical reactions on the assumption that any living organisms contained in the soil sample would breathe—absorb some gases and release others—and thereby reveal themselves. The experiment demonstrated that the soil samples released carbon dioxide and oxygen as soon as they came into contact with the nutrient solution. This looked promising, but the rate at which the reaction proceeded corresponded precisely to that expected from a chemical reaction. The soil samples obviously contain strong peroxides or superoxides, oxygen-rich compounds such as hydrogen peroxide or potassium superoxide that, when in contact with moisture, act like a tablet of sodium bicarbonate dropped into a glass of water.

Taken as a whole, the experiments on the detection of life have suggested the idea that, under the influence of a completely different environment and the virtually unrestricted incidence of ultraviolet light, there might be a chemistry on Mars totally alien to us. Entirely different, extremely bizarre chemical reactions could occur that would not be feasible in terrestrial conditions. Here, again, we come to the conclusion that further expeditions to Mars (by unmanned and manned spacecraft) are necessary to find satisfactory answers to still unresolved questions.

MERCURY—PLANET OF THE INFERNO

The small plant Mercury orbits the sun at a distance of 58 million km, about a third of the distance between our earth and the sun. Its proximity to the sun not only decides the physical conditions on this planet, but also accounts for its poor visibility from earth. In our sky the planet is never more than 28° away from the sun and therefore visible only at certain times shortly before sunrise or shortly after sunset—never in the night sky. Many astronomers never set eyes on it. Nicolas Copernicus was one of them. He is said to have regretted on his deathbed that never in his life had he had the pleasure of seeing Mercury.

Mercury has a diameter of 4,878km—less than a third the diameter of earth. When you consider its size and position in the sky, it is not surprising that—even with the aid of telescopes—it is barely possible to distinguish surface details. For decades we did not even know the speed at which Mercury rotated on its axis. The

Left: This photograph of a recent impact crater on Mercury was taken by Mariner 10 on March 29, 1974, from a distance of 34,000km. The crater has a diameter of 120km.

Right: Taken by Mariner 10 from a distance of 200,000km, this mosaic consists of eighteen photographs and shows two thirds of the surface of Mercury. The largest craters have diameters of 200km.

clues provided by observation through telescope are too indistinct to permit a reliable calculation of Mercury's period of rotation. (Observing the shift of surface details makes this calculation possible for Mars, Jupiter, and Saturn.) The first positive details about the rotation of Mercury became available in 1965, with the aid of radio astronomy. Reflections of the radar beams of a radio telescope—which is not only a receiver but a transmitter—by Mercury showed that the planet takes fifty-nine days to revolve once on its axis. This is two thirds the time it takes for its orbit around the sun. On Mercury the year lasts eighty-eight earth days, and its days pass hardly quicker than the years.

In 1962, when the radio waves emitted by the planet were observed, it was discovered that Mercury does not revolve synchronously. The observed radio waves came not only from the day side of the planet, heated by solar radiation, but also from the night side. This meant that the night side must sometimes be lit and heated by the sun, otherwise no radio waves could be emitted from there. The old claim, going back to Schiaparelli, that Mercury's revolution was synchronous was thereby refuted. More far-reaching information about the sun's nearest neighbor was not acquired until March 1974,when space probe Mariner 10 passed Mercury's surface at a distance of 756km, and sent 2,300 photographs back to earth. Mariner 10 had been launched at Cape Canaveral in November 1973 with the objective of investigating both Venus and Mercury.

The pictures Mariner 10 transmitted were excellent reproductions of Mercury's surface. They showed a planet pockmarked with hundreds of craters and, at first glance, a planet very similar to the moon. Calculations of the orbit had shown that after flying by Mercury Mariner 10 would enter a very eccentric orbit around the sun, which would take it past Mercury again every 176 days. The probe transmitted more pictures and scientific information to earth during its two subsequent encounters with Mercury in September 1974 and in March 1975. After the third fly-by Mariner began to tumble out of orbit; the propellant for its stabilization had been exhausted.

More detailed analyses of the pictures of Mercury revealed that the similarity with the lunar landscape was true only on first approximation. Closer inspection showed that the landscape of Mercury has its own characteristic features, different from those of the moon and presumably because Mercury is larger and more massive.

Whereas the moon's mass is only 1.25 percent that of earth, Mercury's is 5.5 percent. This considerable difference accounts for the fact that the gravitational force on the surface of Mercury is 38 percent of that on earth, whereas on the moon it is only 16 percent. This means that a man who weighs 75kg on earth would weigh only 12kg on the moon, but 28.5kg on Mercury. This difference in gravity obviously affects crater formation. When planetoids or giant meteorites plunged into the moon, huge boulders were ejected and formed secondary craters at great distances from the point of initial impact because of the moon's weak gravitational force. On Mer-

This picture of Mercury—taken by Mariner 10 on March 29, 1974, from a distance of 5,600km—shows an area on the Northern hemisphere of the planet. It is a transitional zone between a terrain that has many craters and one that is relatively flat. The plains on Mercury are probably the result of volcanic activity; the flowing lava filled many craters. The region shown here is about 490km wide.

cury (whose craters are also of meteorite or planetoid origin), the distance between comparable impacts is only one sixth of that on the moon. As a result the craters on Mercury are not so closely spaced, nor do they overlap one another as they do on the moon. There are numerous plains between the craters of Mercury. Another feature exclusive to Mercury are cliffs and mountain ranges hundreds of kilometers long, overlapping embankments, of 1 to 2km in height.

When flying by Mercury Mariner 10 recorded, among other phenomena, the planet's magnetic field. Although its strength is less than 1 percent that of earth, it indicates that Mercury must have a core rich in iron. From the mean density of the planet, the diameter of this iron core was calculated at 3,600km. But because the diameter of Mercury is only about 4,800km, the mantle is no thicker than 640km. The existence of this iron core and the magnetic field it generates are two further differences between Mercury and the earth's moon.

Habitability of Mercury and life on its surface is out of the question. Mariner 10 measured temperatures on both the day and the night sides of Mercury during its fly-by. Day-side temperatures of 427°C and night-side temperatures of −173° were recorded. This enormous difference shows that Mercury has no atmosphere comparable with that of the earth, an atmosphere that would ensure a healthy temperature equalization between the extremes of the heat of the day and the cold of the night. Analyses have detected only an extremely thin atmosphere on Mercury composed mainly of the noble gas helium. Atmospheric pressure on the surface is less than 2 billionths of a millibar, that is, 500 billionths the atmospheric pressure on the earth's surface.

Humans would be unable to survive the surface of this planet for even a few minutes. At the already mentioned temperature of more than 400°C, a glittering sun in area nine times, in diameter three times as large as its disk in the earth's sky, scorches from above. From Mercury the sky is deep black, because the planet's thin atmosphere is unable either to scatter the sun's radiation or to absorb parts of it. In addition to the visible radiation, the solar short-wave ultraviolet rays strike the surface of Mercury at full intensity. Lead, zinc, sulfur, and other substances melt at these temperatures; if rocks on the surface of Mercury contain them, they would ooze out during the day, form silvery puddles, and resolidify at night.

VENUS—PLANET OF THE CLOUDS

To a human visitor Venus—often apostrophized as the Morning or Evening Star—would be a hell similar to Mercury. Venus does have an atmosphere. However, it is so dense that the first space probes that landed on Venus were immediately squashed. The Venusian surface atmospheric pressure is 90 times that of earth. Of this atmosphere, 97 percent consists of carbon dioxide, 2 percent of nitrogen, and the remaining 1 percent represents other gases.

Conditions of identifying Venus from Earth are more favorable than with Mercury. At certain times Venus rises up to three hours before the sun (as Morning Star), at others it sets up to three hours after the sun (as Evening Star). Opportunities for observation are therefore ample. At the times of optimum visibility Venus is far brighter than Sirius, the brightest fixed star in the sky; and next to the sun and moon it is altogether the brightest celestial object. But apart from the phases, comparable with those of the moon, that Venus reveals, no details except occasional blurred, ill-defined, weak spots can be distinguished—even through large telescopes. The planet is shrouded by a blanket of clouds. Its phases were observed by Galileo and together with the apparent change in its size—largest as a narrow crescent, because it is then closest to earth, smallest shortly before "Full Venus"—he adduced these phenomena as visible proof of the heliocentric nature of the universe. But not even the occasional vaguely defined spots have helped modern astronomers in their efforts to determine the planet's period of revolution on its axis.

Before the successes of probes, our information about Venus was restricted to a few bare facts: its dimensions are very similar to those of the earth. At a diameter of 12,228km, it is only 4 percent smaller than the earth; and its mass is 81 percent that of earth. Its distance from the sun, 108 million km, is more than two thirds that of the earth's. These similarities, and the lack of knowledge about physical conditions on Venus, encouraged for a long time the hypothesis that our nearest neighbor among the planets might resemble the earth very closely, indeed that it might harbor living creatures comparable to those on the earth.

The results of later astronomical research, especially with space probes, have refuted these views. In the early 1960s radar scans of the planet established its period of revolution at 243 days retrograde. This means Venus revolves on its axis in the direction opposite that of the revolution of Earth, Mars, Jupiter, and Saturn. On Venus the sun rises in the West and sets in the East. The already mentioned 243-day revolution is the sidereal period of revolution of the planet on its axis, that is the time required for a complete revolution. Because Venus completes its orbit around the sun in 224.7 days, the length of the day in the sense of the period of sunlit day plus dark night is 116.8 earth days. In the equatorial belt of Venus the sun therefore requires about 58 earth days to travel from the point it rises in the West to where it sets in the East.

In December 1962 Mariner 2 (the first successful Venus probe after Soviet and American failures) recorded, as it flew by the cloud-covered planet, temperatures of up to 430°C in the Venusian atmosphere. Later measurements by landers, sent to Venus by the Soviet Union from October 1967 onward, showed surface temperatures of the planet of 475°C, 45°C hotter than Mercury, caused by a "hothouse effect" produced by the carbon-dioxide-rich Venusian atmosphere. The atmosphere of the planet retains heat radiation

This picture of Venus was taken by Mariner 10 on February 6, 1974, from a distance of 720,000km. It shows Venus in ultraviolet light and was the first to reveal structures in the Venusian cloud cover.

very efficiently so that nocturnal cooling is negligible and very little heat escapes into space. Later investigations by American Pioneer landers in December 1978 showed that the rocks on the night side of Venus are so strongly heated during the day that they remain red hot.

The cloud cover was found by the Soviet probes Venera 7 and Venera 8 (December 1970 and July 1972) to begin at an altitude of 65km above the surface of the planet, but does not extend below 35km, where the atmosphere is relatively clear and transparent as far as the enormous density of the bottom layers allows. Atmospheric pressure on the surface of Venus is as high as the pressure on earth at a depth of 900m of water.

Four American Pioneer landers arrived on Venus on December 5, 1978, carried out numerous measurements during their descent, and radioed information back to earth even after the relatively soft landing. They were able to distinguish three well-defined cloud layers between 48km and 65km altitude, but found a layer of haze between 31km and 48km. According to the data the atmosphere is 94.4 percent carbon dioxide, 3.4 percent molar oxygen, and 0.14 percent water. Two hundred parts per million—equivalent to 0.02 percent—consist of sulfur dioxide.

The clouds of the Venusian atmosphere are obviously the result of violent chemical reactions between hydrogen sulfide and sulfur oxides. They probably consist mainly of oxygen, water vapor and sulfur compounds. Presumably sulfuric acid is also present in the clouds. The thought that rain could consist of pure sulfuric acid is one more uninviting aspect of this planet.

Mariner 10, on its flight by Venus in February 1974, found that the Venusian atmosphere rotates at considerable speed. The upper atmosphere rotates around the planet once every four days, in a retrograde direction. From the difference between the rotation periods of the sphere of the planet and of its atmosphere we can deduce wind speeds of 100m/sec or about 360kmph. However, Venera 9 and Venera 10 discovered, in October 1975, that the winds do not occur on the surface of the planet; they measured surface wind speeds of only 3.5m/sec, or 12kmph. The calculated differences in speed are absorbed by wind shear at higher altitudes. In this context one should remember that in the earth's atmosphere, too, speeds of more than 500kmph have been recorded in the jet streams at altitudes between 12 and 15km.

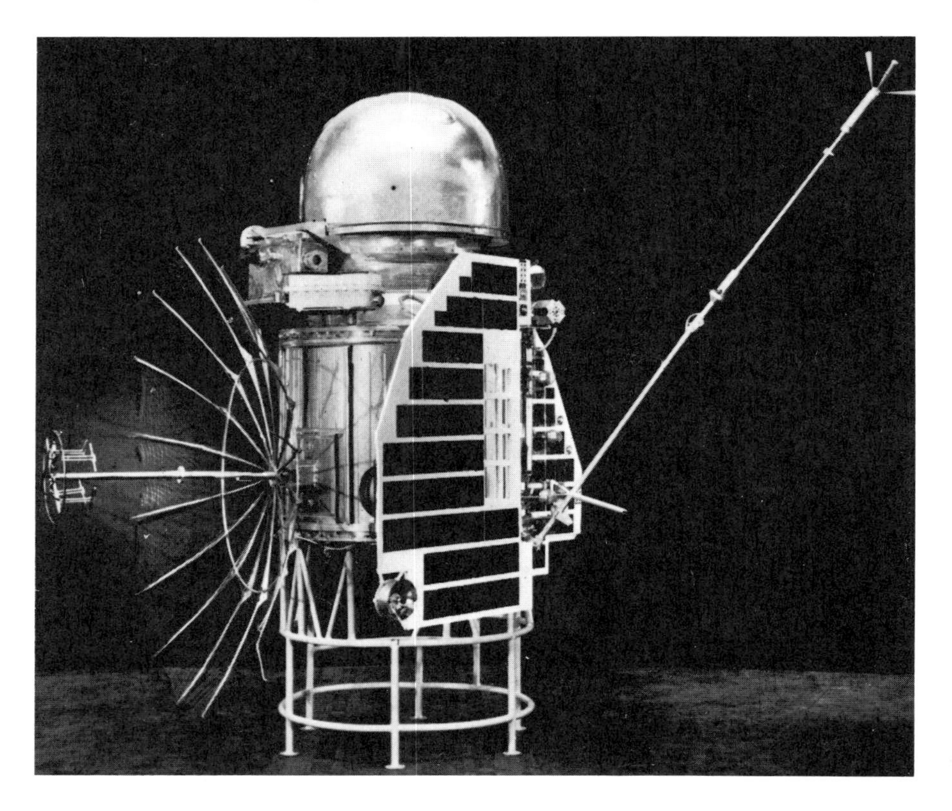

The Soviet probe Venera 10 landed on Venus on October 25, 1979, and transmitted photographs and data. Information gathered by this and other probes revealed Venus as one of the most inhospitable places in the solar system.

The most recent radar measurements of Venus have shown huge canyons, mountain ranges, and craters with diameters of more than 300km. Measurements by the 1978 Pioneer Venus probe have contributed to our present mapping of 83 percent of the Venusian surface between 75° latitude North and 63° latitude South. More than half of the area between these limits is very flat; the maximum difference in height is no more than 100m. But there are two huge highlands and many craters, probably of meteoritic origin. The craters measure between 200 and 400km across, but are only 200 to 700m deep. There are indications that many of them have central peaks. On Ishtar Terra, one of the highlands, the Maxwell Montes (the highest mountains so far discovered on Venus) rear up to a height of 9,000m above their surroundings. Measured from the defined zero level of Venus (a kind of artificially determined "sea level") their height is 11,800m. The Maxwell Montes are the roughest terrain on Venus.

Ishtar Terra is about the same size as Australia. The other highland of Venus, Aphrodite Terra, is approximately as large as northern Africa. The terrain is rough, ravined, and rock-strewn, as is Ishtar Terra; but its mountains are not quite as high.

There are numerous parallels to the properties of the surface of the earth and of Mars; evolutionary comparisons are also possible. The geological development of Mars and Venus, for instance, was perhaps different from that of the earth. The gigantic valleys on the two planets, much deeper than comparable formations on the earth, presumably were caused by geological faults. Thus, they were formed by genuine movements and fractures in the crust of these planets and not, as on earth, by erosion.

The Pioneer probe found on Venus the largest canyon we know of in the solar system. It is a rift valley 4,500m deep, 280km wide, and 1,500km long. The appearance of this canyon points to internal forces that must have broken up the crust of the planet and created this valley formation; neither erosion nor the impact of a meteorite would produce such a long, straight-line feature.

After Mariner 10 had recorded the very high wind speeds in February 1974, it was assumed that the surface of Venus must have been considerably flattened by erosion. But the October 1975 pictures transmitted by Venera 9 and Venera 10 showed many pieces of rock lying close to the landing sites. The wind speeds measured on the surface of Venus explain the apparent contradiction: the slow wind speeds near the surface are incapable of producing major erosion effects. Infrared observations above the poles of Venus by Pioneer indicated almost circular downwind regions measuring more than 1,000km across; these may play an important part in the overall atmospheric circulation of Venus. And above the North Pole Pioneer discovered a hole with a diameter of 1,100km. This "hole" is completely cloudless or at most contains only occasional clouds. It is assumed that the surrounding clouds in this downwind zone are carried to lower, warmer altitudes, where they evaporate and disappear. These processes could be the key to understanding the large-scale atmospheric circulation of Venus. The measurements of the four landers have shown that the temperature and pressure data are the same in 99 percent of the Venusian atmosphere on the entire planet—on the poles, on the equator, on the day side, and on the night side. The atmospheric distribution system must be highly efficient to produce such a result.

Characteristic dark markings in the form of the letters C and Y spanning the entire sphere of the planet and visible only on ultraviolet photographs, are either huge openings in the white, strongly reflecting upper cloud layer or violent upwind zones of sulfur compounds extending to altitudes where they become visible through the thin uppermost cloud layer. The ultraviolet spectrometer on Pioneer Venus identified sulfur dioxide clouds whose distribution corresponds to the dark C- and Y-shaped phenomena. Venus may therefore have an atmospheric sulfur circulation that would correspond to the terrestrial counterpart of water circulation.

Although details of such a circulation are not yet known, it could consist of "raindrops" of sulfuric acid falling through the clouds. They become hotter as they reach the lower layers until at the bottom boundary of the densest cloud layer, an altitude about 48km above the surface of Venus, they evaporate and split up into their components. The resulting products—water, sulfur dioxide, oxygen, and a number of sulfur compounds—rise again and reform in the upper atmosphere into sulfuric acid with the aid of the ultraviolet radiation of the sun.

A few years ago the Venera probes recorded lightning flashes in the planet's atmosphere at altitudes between 32 and 2km. These flashes, confirmed by Pioneer probes, occur at surprising frequency, up to 25 per second. So rapid is this frequency that the human eye cannot distinguish the individual flashes; they would be perceived as a constant light, presumably accompanied by continuous roaring thunder. During earlier telescopic observations of

---

These two representations of the surface of Venus were drawn on the basis of radar data transmitted in 1979–80 by Pioneer-Venus. The top picture shows Ishtar Terra, a Venusian highland the size of the United States. The highest mountain on Venus, Maxwell Montes, rises 10,600m above the defined "sea level." It takes up a large part of the eastern side of the highland. The dark, round structure on the western flank has a depth of over 900m and could be the crater of a volcano. The bottom picture shows a gorge in Aphrodite Terra, the largest highland on Venus. It is 4,800m deep; 280km wide; and 2,250km wide.

the night side of Venus, an ashen-gray light was occasionally seen; some observers suggested this light was the result of lightning discharges. The information from the Pioneer and Venera probes confirms this assumption. Venera 11 and Venera 12 were able to determine the extent and distance of the thunderstorm zones. One storm recorded by Venera 11, for instance, occurred at a distance of about 1,500km from the landing site of the probe, and it measured about 150km across.

The Pioneer probe also discovered strange magnetic fields in the ionosphere, the electrically conductive layers of the Venusian atmosphere. Peculiar tangled lines of magnetic fields were demon-

Contour map of Venus based on radar measurements from Pioneer-Venus 1.

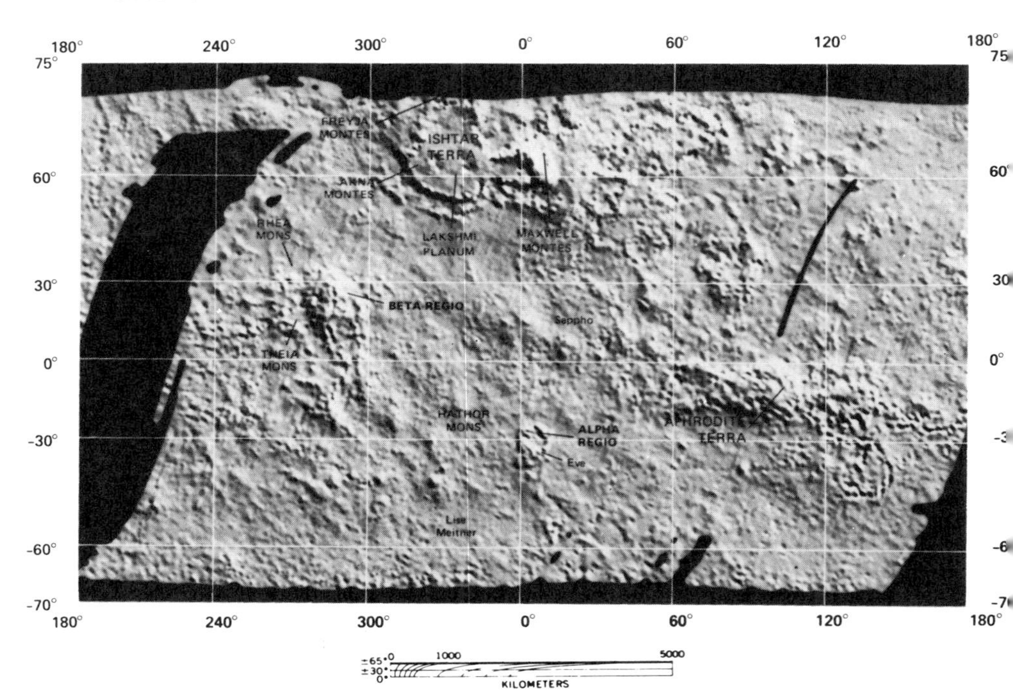

As of May 1980 Pioneer had not yet transmitted data for the areas in black. No human observer and no camera has so far been able to survey this because of the dense Venusian cloud cover. Less than 1 percent of the planet's surface was known before the measurements sent by Pioneer.

strated. Because Venus has no magnetic field of its own, the lines of magnetic force are thought to be caused by solar wind. This stream of protons, atomic nuclei of helium, and electrons constantly bombards the ionosphere of Venus. During impact the particles of solar wind are deflected and most of them flow round the planet. At the same time the strength of the magnetic field is reinforced where this collision occurs. The possibility exists that this magnetic field in the ionosphere spreads in the form of tangled, ropelike lines of force.

Although the phenomena discovered on Venus are most impressive and instructive, they leave many questions unanswered. Several NASA scientists are, therefore, in favor of another Venus project. The goal of a new project is a precise mapping of the surface with the aid of radar. The project has been named VOIR, an acronym for Venus Orbiting Imaging Radar (also French for "to see"). In VOIR radar will be used to "see" the surface of Venus. Its rays readily penetrate the opaque atmosphere, and their reflected signals will transmit information about the structure of the surface and its relative elevations. No matter how interesting the information gained so far about the surface of Venus, all we have is a rough map of the terrain of the planet. VOIR is intended to cover 70 percent of the surface of Venus at a resolving power of 600m and to produce maps of about 2 percent of it showing features measuring no more than 150m across.

In addition, VOIR would conduct investigations about the gravitational field, the atmosphere, and venerophysical (corresponding to geophysical) questions of the planet. Measurements to check the Theory of Relativity are also planned.

December 1984 would offer a favorable opportunity from the aspect of celestial mechanics for launching this VOIR probe. If this deadline is met (it depends on the timely availability of funds for the project), the probe will arrive on Venus in May 1985. Its launch, by the way, would not be by conventional rocket but by the new reusable space shuttle. The shuttle would inject the probe into orbit around the earth where it would be ejected from the cargo bay of the shuttle and blasted by a rocket stage into the trajectory toward Venus.

Five months later the probe would arrive near Venus, where it would be injected into an elliptical orbit around the planet. This would bring it periodically down to a distance of 300km from the surface of Venus and out to a distance of 19,000km. During this orbit the gravitational field measurements would be taken. Two months later this elliptical orbit would be converted into a circular one 300km above Venus. Only from the circular orbit could the planet be mapped with the imaging radar equipment.

Such an undertaking would yield much information about the atmosphere of Venus, the circulation of its atmosphere, the tectonics of its surface, volcanic activity in the past and perhaps in the present, and erosion by wind and by water. This is also essential to a better understanding of our own planet and of its role in the planetary system.

JUPITER—THE GIANT PLANET

Jupiter, the largest planet in our solar system is over five times as distant from the sun as is earth—5.2 astronomical units or 778 million kilometers separate it from the sun. The light ray, which travels 300,000km every second, takes 1.25 seconds to travel from the moon to the earth, 8.3 minutes from the sun to the earth; it takes 43 minutes for light to travel from the sun to Jupiter. This remote region of the solar system is icy cold, and the temperatures of the upper atmosphere of Jupiter, measured with astronomical methods from earth and by space probes, are in the neighborhood of $-130°C$.

Jupiter is an enormous sphere of liquids, gases, and particles of ice. Its diameter is more than eleven times the earth's; the equatorial diameter of the planet is 142,700km. This means that it would require 1,400 spheres the size of the earth to fill Jupiter if it were hollow. Jupiter's mass is 318 times greater than the earth's, and 2.5 times more massive than all the other planets, asteroids, meteorites, and comets of the entire solar system added together. Nevertheless, Jupiter revolves faster on its axis than all the other planets—9 hours 55 minutes for a complete revolution. The equatorial belt revolves a little faster still, taking 9 hours 50 minutes.

Jupiter does not have a solid body. What we see of it from earth is its dense envelope of clouds. It can be observed very well even with a small- or medium-power telescope. It was the continuous observation of the planet's cloud structure that made the amateur astronomers' contribution to its exploration so important. Thousands of drawings of the planet, most of them by amateurs, have been produced.

Below the planet's cloud cover, at a depth of about 1,000km, the Jovian atmosphere gradually merges into a "soup" consisting mainly of liquid hydrogen. Jupiter is composed of 82 percent hydrogen and 17 percent helium—all other elements and compounds account for no more than 1 percent. The mantle of liquid hydrogen probably extends to about 25,000km below the cloud cover. Obviously pressure increases steeply with increasing downward from the cloud cover; at the depth of 25,000km it is probably more than 3 million atmospheres, where, because of this high pressure, the liquid hydrogen is converted into the metallic state. Jupiter may also have a core, whose diameter is thought to be 12,000km, about the same as the earth's.

The peculiar structure and striking composition of this planet is, again, based on the historical development of the planetary system. We have already pointed out that the site of origin in the solar system and the temperatures prevailing there at the time were decisive to the cosmogony of the planets. Because of the extremely low temperature owing to the distance between Jupiter and the sun, the interplanetary dust present at the beginning of the formation of the planets was largely covered with ice. These ice granules probably collapsed on each other at enormous speed, forming as it were the foundation stone for the evolution of Jupiter. Detailed calculations and theories exist about this question, but only the most important ones will be covered here. Many can be rendered in only mathematical terms. Nevertheless, we can outline an approximate picture of the processes that took place.

After the consolidation of the ice-coated dust particles into a core, a phase must have been entered during which—due to low temperature in space and the gravitational force of Jupiter's core—atoms of hydrogen and helium were captured by this core. In fact, an unstable state must have become established in the vicinity of Jupiter's core which, after reaching critical minimum size, led to all the hydrogen and helium nearby rushing to the new planet and combining with it. The result was a giant composition that corresponded to that of the interplanetary gas and dust in the system.

Because of its rapid revolution Jupiter's gas and liquid sphere exhibits appreciable oblateness or flattening. The planet's equatorial diameter, 142,700km, is more than 9,000km larger than its diameter from pole to pole. It can be observed quite clearly in the telescope; instead of a round disk we see an elliptical one elongated in the direction of the clouds. Jupiter is not a sphere, but a typical so-called ellipsoid of revolution. The earth, too, is flattened, but considerably less because of its smaller dimensions and slow revolution. Its equatorial diameter exceeds the polar one by only 42.92km.

These figures are, of course, approximate because both Jupiter and Earth deviate from the ideal spherical shape in other respects. The extent of these deviations is small, but modern satellite technology is capable of detecting them. The second American earth satellite, Vanguard 1, found as long ago as 1957 that the Northern and Southern hemisphere of the earth are asymmetrical with respect to each other; the Northern hemisphere is a little "slimmer" than the Southern. This led to the exaggerated statement that the earth is pear-shaped. But the difference is no more than 15m, and therefore exclusively of theoretical interest.

The rapid revolution not only results in Jupiter's oblateness, but also affects the appearance of the planet. What we see of Jupiter in the telescope is only the cloud cover. But unlike Venus, whose layers of cloud have a structureless appearance in visible light, Jupiter's show a great many details and structural features. The first features that strike the observer looking at Jupiter through the telescope are several light zones and dark belts. They are so regular that the individual zones and belts were identified and given names: for instance, SEB (the Southern Equatorial Belt), NEZ (the Northern Equatorial Zone), NPB (the Northern Polar Belt), and STZ (the Southern Temperate Zone). In addition to belts and zones many details such as notches, large and small bridges, garlands, rodlets can be observed in the atmosphere. Some of these features are persistent, others are visible only for a few months or years.

One of the most interesting elements, first observed by the astronomer Giovanni Domenico Cassini in 1665 and seen at fluctuating intensity ever since, is the Great Red Spot. This is an oval reddish feature in the SEZ, partially projecting into the SEB, and whose nature was obscure for a long time. The Great Red Spot's major axis is 30,000 to 40,000km; its minor axis, 14,000km, about 1,200km longer than the diameter of the earth. Three spheres the

This mosaic of Jupiter—sent by Voyager 1 on February 26, 1979—consists of nine photographs and was taken from a distance of 7.8 million kilometers. The Great Red Spot can be seen on the right. The smallest discernible objects in the picture measure 140km. The cloud structures, even within the Great Red Spot, reveal the great complexity of the atmospheric processes on this planet. The clouds, for instance, flow around the Red Spot so that the friction generated is very low.

size of the earth could be comfortably accommodated along the major axis.

In Plassmann's *Astronomy* (published in 1913) we read about this Great Red Spot: "This structure attracted attention for several decades, especially during the Eighties. It seems to be a centre of effusion extending into great depths." Its intensity has changed in the course of time; in the past this object displayed a far more luminous red hue than at present. Mainly from 1878 to the end of the last century it was a brilliant, red, prominent feature, but since then both its color and its prominence have declined considerably.

Some observers suspected that the Great Red Spot was a huge lump of frozen helium; others interpreted it as eruptions from the interior of the plant; some saw it as a part of the solid surface of Jupiter; still others thought it was a solid floe of enormous dimensions floating on deeper, probably liquid or viscous layers, with a base supported by a solid stratum. But its buoyancy—according to Littrow and Stumpff in *Magic of the Sky*—would be so strong that even weak forces generated by currents would be enough to lift it off its support and allow it to drift in the current. Since its discovery it has been lapped by System II (the rotary system of the temperate latitudes, revolving once every nine hours fifty-five minutes already by several revolutions of the planet). It can, incidentally, hardly be assumed that the drifting floe, regarded as the most likely explanation of the Great Red Spot, will become visible. Rather, the phenomenon is probably due to vertical currents in the atmosphere, induced by the presence of such an enormous body and pushing upward along its edges.

It was left to Pioneer 10 and Pioneer 11, which flew by Jupiter in December 1973 and December 1974, and especially to the two Voyager probes, which passed Jupiter in March and July 1979, to solve the riddle of the Great Red Spot and of many other Jovian events. What we see there is a kind of weather compared with which our terrestrial counterpart is indeed modest. The measurements and the pictures the four probes transmitted to earth made it clear that the bright zones of the planet are high-pressure areas, anticyclones, and the dark belts are atmospheric low-pressure areas, depressions. In the zones warm gases rise, cooling as they reach the upper limit of the clouds. These cool gases flow sideways down into the dark low-pressure areas. As a result of the rapid revolution of

Jupiter, violent jetstreams and trade winds occur along the boundaries between zones and belts. It is a meteorological system of currents which we would also have on earth were our planet to revolve more quickly: the high- and low-pressure areas both extend around the entire sphere of the planet. This sytem of linear high- and low-pressure areas continues far into the north and south; only in the polar regions do the large-scale orderly currents become enormous cyclones. The pictures produced by the Pioneer and Voyager probes show excellent examples of this.

Plassmann suspected the truth about the Great Red Spot: it is indeed a "hearth of effusion extending to great depths"; it is a huge high-pressure area in the southern zone, which is already under high pressure; it is, as it were, an ultra-high-pressure zone. As the high-pressure zones protrude beyond the low-pressure belts, the Great Red Spot protrudes beyond the high-pressure areas of Jupiter. Infrared photographs taken from the earth and infrared measurements by Voyager show that the temperature of the Great Red Spot is lower than that of its surroundings; they thereby confirm its projection above the rest of the planet's clouds. It is an extremely stable formation that has been in existence for more than three hundred years. This stability is due to the fact that the huge cyclone rotates in itself and generates little friction with the masses of gas surrounding it. The entire Red Spot revolves counterclockwise on its axis once in six terrestrial days. It is embedded in its surroundings so that hardly any friction occurs in its marginal areas, because the current of the tropical zone moves —as seen from the Spot—against the current in the Southern Equatorial Belt. The Spot thus "rolls," like a ball between two planes moving in opposite directions. If the Great Red Spot were smaller the forces of friction would be greater and it would have long ago ceased to exist. To the south of the Great Red Spot there is, for example, a smaller, oval storm center, which has been there for only about forty years. Many other small features come to light, only to disappear after one to two years. Their internal structure corresponds to that of the Great Red Spot.

In the northern areas of the planet Pioneer and Voyager found several brownish ovals, probably openings in the upper cloud cover, that afford a view of the deeper, warmer cloud layers. These objects persist for periods of one to two years. The pictures obtained by the Voyager probes show that the storms raging on Jupiter are far more complex than they appear from earth. Many turbulent movements occur within the belts and zones.

Between January and August 1979 Voyager 1 and Voyager 2 transmitted more than 33,000 photographs of Jupiter and its five largest satellites. The analysis of this material has provided us with much insight into the dynamic processes in Jupiter's atmosphere; we now have a much clearer idea of it than before. We are able to deduce from this information that the objects we observe in the Jovian atmosphere all move at the same speed irrespective of their size. This indicates that the observed events represent genuine movements of matter rather than the propagation of wave energy through relatively stationary matter. The observed rapid brightening of various white regions, which spread after the increase in their brightness, points to violent vertical convection currents. The Voyager pictures have refuted the previous assumption that the movements observed in the polar regions of Jupiter are completely chaotic. There, too, predominantly east-west winds, following the pattern of the belts and zones, were recorded.

Violent thunderstorms with blinding lightning flashes also rage on Jupiter. They are even more extensive and intense than those on Venus and probably also occur at the upper boundary of the cloud layer. Polar lights, too, were photographed on Jupiter by the cameras of the Voyager probes. These luminous phenomena appear in ultraviolet and in visible light. They seem to be linked with volcanic eruptions on the Jovian moon Io. The two Voyager probes also determined the temperature and atmospheric pressure at various altitudes above Jupiter. In the upper regions a gas pressure of 5 to 10 millibar was found, corresponding to 1/200 to 1/100 of the atmospheric pressure at ground level on earth. The temperature measured in this region was −113°C. Also, an atmospheric phenomenon well known in terrestrial meteorology was recorded on Jupiter: an inversion layer, in which the normal temperature gradient—the higher the colder—is inverted, warmer air overlying colder layers. This warmer layer of gas was detected in the Jovian atmosphere within the range of 35 millibar gas pressure, that is, at a depth of about 90km.

Jupiter also possesses an ionosphere. In Jupiter's ionosphere, as in that of earth, temperatures are very high; as a result atoms and molecules move at very high speeds. The temperature in the Jovian ionosphere was recorded by Voyager at 830°C. Further information confirms that the temperature of the ionosphere changes very greatly in time and space. Voyager determined the proportion of helium in the Jovian atmosphere as 11 percent of the hydrogen proportion, a value significant in cosmogonic considerations not only in conjunction with the development of Jupiter, but also of the solar system as a whole.

Still unresolved are questions of color: What causes the colors of the clouds on Jupiter? Why are the belts dark and of different colors? the clouds white or gray? What gives the Great Red Spot its

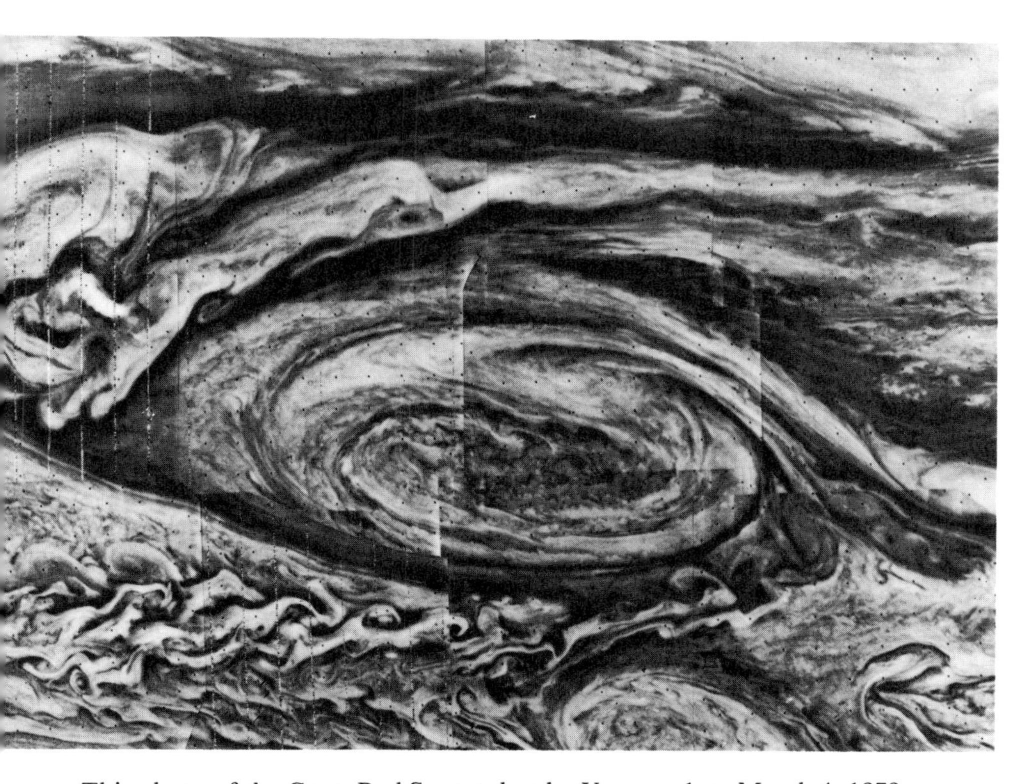

This photo of the Great Red Spot, taken by Voyager 1 on March 4, 1979, from a distance of 1.8 million kilometers, reveals the details of this mighty anticyclone. The Great Red Spot rotates counterclockwise (shown by the shape of the clouds along its edge); other detailed features rotate more slowly or even in the opposite direction.

color and why does it change in the course of time? Two theories attempt to explain this. One is based on the assumption that there are substances, as yet undetected in Jupiter's atmosphere, which pass through ultraviolet radiation at the upper boundary of the cloud cover and enter into photochemical reactions, acquiring a rust-brown hue in the process. The components in the Jovian atmosphere known to us—hydrogen, helium, methane, ammonia, and water vapor—are of course colorless and cannot be the cause of the colorations. However, rising ammonia gas crystallizes into white ammonia snowflakes at the low temperature of -130 °C at the upper boundary of the cloud cover; this explains the white of the belts.

The second hypothesis about the colors on Jupiter assumes that the discolorations indicate biological activities. If this theory is correct, Jupiter could be the second planet in our solar system that sustains life, albeit only in primitive form. These considerations are based on a theory suggested by Harold Urey and on experiments performed with Stanley Miller back in the 1950s, when the two scientists were engaged by the question of the origin and the development processes of life. They produced a gas mixture of ammonia, methane, water vapor, and hydrogen—an artificial atmosphere they assumed roughly equivalent to the primeval atmosphere of earth. As far as we know, the chemical composition of this atmosphere is similar to that of the present atmosphere of Jupiter. The two scientists conducted spark discharges in this simulated proto-atmosphere to initiate lightning flashes similar to ones that must have raged during the horrific thunderstorms in the early atmosphere of the earth. They continued this experiment for several days, then examined the chemical composition of their primeval atmosphere. They were astonished to find amino acids—the molecular building bricks of the proteins and precursors of life. These experiments were repeated in many laboratories within the last few years.

In the course of Urey's and Miller's experiments the original glass-clear solution below the artificial proto-atmosphere turned into a turbid, brown soup—of the tone and colors we observe in the belts and in other objects on Jupiter. We knew of the existence of fierce thunderstorms and lightning flashes on Jupiter even before we received photographic evidence of such events from Voyager; it was deduced from radio-astronomical investigations as early as

1955. Intermittent high-energy radio bursts were recorded against the constant background radiation. These crackling noises had obviously been caused by cloud-to-cloud lightning flashes on Jupiter. The possibility cannot be ruled out that during the long time these phenomena have been occurring amino acids (which cause the rust-brown hue) have not only formed, but even developed into living organisms. But at this time it is impossible to prove.

The assumption arising from infrared observations from earth that Jupiter radiates more heat than it receives from the sun has been confirmed by Pioneer 10. The planet radiates about twice as much heat into space as it absorbs from the sun. This phenomenon has not yet been definitively explained. Presumably it is still based on the heat generated during the formation of the planet more than 4 billion years ago. The heat store produced in the course of this fiery process is very great in Jupiter's huge sphere, and cooling very gradual. Another explanation is fractionation; the additional heat radiation of the planet results from the separation of hydrogen and helium inside Jupiter. The giant planet may be "settling," or shrinking. Shrinkage of only 1mm per year could explain the surplus heat radiated by Jupiter. In addition, the thermal energy conveyed from the interior, eventually into the atmosphere, and radiated into space creates weather conditions induced by Jupiter itself in an atmosphere which unlike that of the inner planets is not in radiation balance. This may be the reason for many features in the gaseous envelope of Jupiter which are peculiar when compared with those in the atmosphere of the earth.

Be that as it may, the temperature of the Jovian atmosphere increases with decreasing distance from the planet because of internal heat radiation. Even a few hundred kilometers below the upper cloud boundary temperatures and gas pressures acceptable to terrestrial life are found; primitive single-celled organisms drifting between the clouds would find the environment reasonably congenial. Anaerobic bacteria would not only be able to live in Jovian conditions but would demand them. To the anaerobics, oxygen is as much of a poison as carbon dioxide or methane is to aerobic creatures. On earth anaerobic bacteria are found in hot sulfur springs and at the bottom of the sea. Single-celled organisms were discovered in the hot springs of Yellowstone National Park in 1977; they absorb carbon dioxide, hydrogen, and water and secrete methane gas. It is possible that these organisms, living in strict isolation from oxygen, represent the primeval form from which all life on earth (and perhaps also on some other planets) has developed. Some of the methane recorded on Jupiter may indeed have been produced by such single-celled creatures. Jupiter could well be in the same state as the earth was 4 billion years ago.

Space research with satellites and space probes has taught us that we must not, as we had in the past, regard celestial bodies as self-contained systems. There was the planet in a universe that, apart from some dust particles and atoms, was thought to be absolutely empty. Except that it received light from the sun it was accorded hardly any direct relations with its celestial environment.

We know now that the exact opposite is true. Space is by no means empty. Numerous interrelations exist between events in it and the planets; they are not at all confined to the effects of the sun on the planets; complex interactions proceed between causes and effects of the most varied phenomena. Radiations and fields of force, charged particles and electromagnetic relations, exist in space. The environment of a planet and the phenomena occurring in it can decide the events on its surface and in its gaseous envelope. The earth and its interactions with the universe provide a good example, but the planet Jupiter offers a better one because of the gigantic dimensions of the reactions taking place between it and the space surrounding it.

Jupiter, like the earth, has a magnetic field. The earth's magnetic field is set up by the core of molten iron, which acts like an electric dynamo. The revolving earth generates powerful currents, which in turn produce its magnetic field. Jupiter has no iron core, but the pressure in its interior is so enormous that it converts the liquid hydrogen there into a metallic state. There is no metallic hydrogen anywhere on earth. It is the postulate of a theoretical prediction of atomic physics but does not exist on earth because there is no pressure high enough to convert it into metal. Even our best-equipped laboratories are unable to produce metallic hydrogen, not even for demonstration purposes. But considerations of atomic physics most strongly suggest its existence in the conditions of pressure prevailing in the interior of Jupiter, where billions of tons of gas are supported by the core. The electric currents flowing in the metallic hydrogen as a result of Jupiter's revolution must be huge. Accordingly, the magnetic field is also of gigantic proportions.

First hints of the existence of a magnetic field around Jupiter were provided by radio astronomy during the 1950s, when the planet was found to be, after the sun, our solar system's most powerful radio emitter. The constant radio signals could be explained only by the assumption that streams of electrons move at high speed in a magnetic field surrounding Jupiter. During the flybys of Pioneer 10 and Pioneer 11, its magnetic field was recorded by magnetometers; this gave us our first idea of the intensity, the dimensions, and the interaction of the magnetic field with other phenomena.

Jupiter's magnetic field is enormous. It sometimes extends toward the sun through 100 radii of the planet, that is, to a distance of about 7 million kilometers. It is, however, very powerfully influenced by the solar wind, which, even in the region of Jupiter, bombards at velocities of 450 km/sec. When the strength of the solar wind increases, it compresses Jupiter's magnetic field so that it extends only 3.5 million kilometers toward the sun. To compensate for this, it forms a huge tail to leeward. Pioneer 10 recorded the planet's magnetic tail beyond the orbit of Saturn, a distance of 650 million kilometers from Jupiter. The Jovian magnetic field, incidentally, is the opposite of the earth's; a compass on Jupiter would point not to magnetic North but to the South Pole.

Like the earth (and probably like every planet that has a magnetic field), Jupiter has radiation belts—zones in which charged particles at high altitudes commute between the magnetic North Pole and South Pole as captives of the magnetic field. The inner radiation belt of the earth was discovered with the aid of Explorer 1 at the beginning of 1958; Jupiter's was pinpointed by Pioneer 10 and Pioneer 11.

Their measurements showed that the radiation belt of Jupiter is very large, corresponding to the dimensions of the magnetic field. Jupiter's magnetosphere—the region surrounding the planet in which the magnetic forces make themselves felt—has a more complex structure than the earth's. It consists of two parts. The outer magnetosphere extends from 20 to 100 Jupiter radii (1.4 to 7.1 million kilometers) into space. At the distance of about 15 Jupiter radii it is almost disk-shaped, largely confined to the region above the planet's magnetic equator. Voyager 1 noted that the closed magnetic field lines on Jupiter's far side from the sun lose themselves in a magnetic tail. In the outer magnetosphere the field lines are therefore open.

Such a magnetic tail had been suspected earlier but Voyager 1 was the first probe to investigate it in detail. Voyager 2 succeeded in recording that this magnetic-field tail of Jupiter extends 300 to 400 Jupiter radii—21 to 28 million kilometers—into space, along the supersonic jet stream of the solar wind. It, too, was traced beyond the orbit of Saturn.

Among the various phenomena found in the vicinity of Jupiter are numerous electromagnetic interactions. They include the radiation belts. The inner radiation belt does not extend beyond a distance of about 12 radii (856,000km) from Jupiter. The transition region to the outer belt extends to 20 Jupiter radii. The inner radiation belt contains very many high-energy particles. In the outer belt many particles of lower energy have been demonstrated. The probes recorded intensity fluctuations of the outer radiation belt of a period of ten hours—corresponding to Jupiter's period of revolution on its axis. Altogether, considerable fluctuations and changes are observed in the charged particles in the course of time.

It has also been shown that Jupiter's magnetic field, unlike the earth's, is not generated by a single electric ring current deep inside the planet; there may be several such currents. This is the only explanation for the distribution of high-energy particles around Jupiter. The intense energy bursts in the long-wave radio emission of Jupiter can also be interpreted only on this basis.

One of the most interesting discoveries is a ring of dust surrounding the planet. Until recently the only ring known to us was that of Saturn. (In 1977 a system of rings was discovered around Uranus.) The Jupiter ring is not as conspicuous as that of Saturn; it is invisible from earth. The two Voyager probes discovered its presence. It consists of material orbiting the planet about 57,000km above the cloud cover. Its thickness is 10km, its width 6,500km, and it has a well-defined outer boundary. The inner boundary merges into a haze-filled region, which extends to the Jovian cloud cover.

Several scientists—from the Max Planck Institute of Nuclear Physics in Heidelberg and from the Jet Propulsion Laboratory in Pasadena—have formed a hypothesis about the origin of Jupiter's ring. They point to Io, one of Jupiter's many moons. Io emits volcanic dust into space. This dust collides with other, larger bodies (perhaps unidentified moons) and knocks particles out of them. These particles get caught in the combined gravitational fields of the planet and its moons and so form the ring.

JUPITER'S MOONS

Jupiter has, as far as we know, sixteen moons. Experience has shown that our count may well be incorrect. During the last several years, three more Jovian moons were discovered on photographs of Jupiter taken by the Voyager probes. Voyager 1 and Voyager 2 were to concentrate on the Galilean satellites—the four largest and earliest-known of the moons. They also investigated Jupiter's innermost moon, Almathea.

Almathea

Almathea, a small red moon, orbits the planet at a distance of 2.54 Jupiter radii—181,000km from the center of the planet, almost 110,000km from the cloud cover. (Recall that our moon is at a distance of 59 earth radii, 378,000km, from the earth's surface.) Almathea is not spherical, but elongated. Its major axis measures 170km; its minor one, 130km. This tiny moon takes only twelve hours to orbit the planet. Because of its small size, Almathea can have no atmosphere.

An interesting question—one that cannot yet be answered—is whether Almathea's striking color is a characteristic of the material of which this moon consists or if it is due to changes in the surface material or due to some kind of precipitation. Voyager photographed Almathea from a distance of 425,000km. The smallest objects that can be distinguished on the photographs (unfortunately not very sharp) are about 8m in length. The impression given by the pictures is that the moon has suffered from violent bombardment; some details of the surface look like craters. This bombardment may have produced Almathea's odd shape.

Io

Far more interesting than the data about Almathea is that about the Galilean satellites. The innermost is Io. Its distance from the planet is 442,000km or 6.19 Jupiter radii, and it takes 1.8 days to orbit Jupiter. At a diameter of 3,640kms, Io is almost 5 percent larger than the earth's moon.

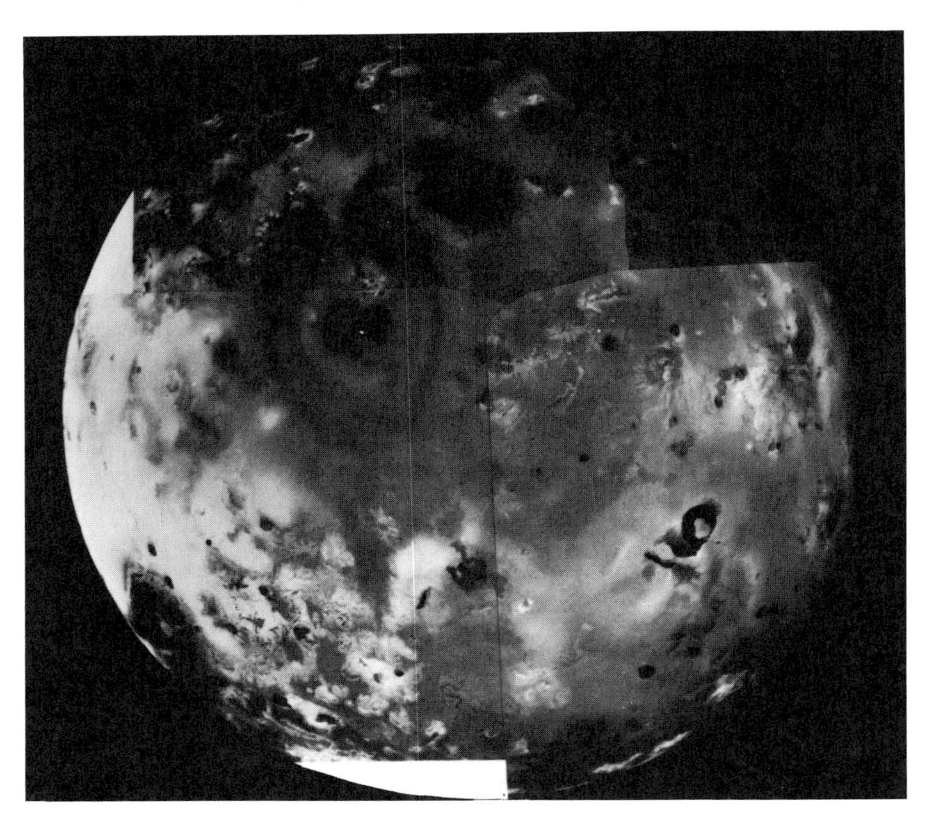

Io, photographed by Voyager 1 from a distance of 376,951km, is one of the most interesting satellites in the solar system. Apart from the earth it is the only known body in space with active volcanoes. Everything points to Io being of recent origin, and many details of its surface are probably of volcanic or at least internal origin. The surface is extensively covered with sulfur, sulfur compounds, and other salts.

The photographs of Io show a very colorful satellite with red, yellow, orange, brown, blue, black, and white hues competing with each other on its surface, with red and yellow dominant. The bright colors and spectrum analysis indicate that the material covering Io's surface probably consists of solid sulfur, frozen sulfur dioxide, and other sulfurous substances. Sulfur frequently enters into complex compounds characterised by the gorgeous colors we have found on Io. Io caused a sensation in the astronomical world when Linda Morabito, a technician at the Jet Propulsion Laboratory, discovered active volcanoes on it as she analyzed the pictures transmitted by Voyager 1. This makes Io the only body in the solar system, aside from earth, on which volcanoes are active today.

What Morabito found looked at first like a cloud of smoke, comparable to a huge umbrella curving above part of the satellite's sphere. Further study indicated the presence of an active volcano. All other photographs transmitted by Voyager were then carefully scrutinized for other evidence. As a result eight volcanoes were found in the most diverse regions of Io's surface. Four months later, when Voyager 2 reached the Jovian system, six volcanoes were still active.

Measurements showed that the volcanoes usually eject clouds of smoke up to an altitude of 100km; however, in one case an altitude of 300km was recorded. Their sulfurous material is issued at a speed of 1km/sec. This rate of eruption is far higher than that of terrestrial volcanoes such as Etna or Vesuvius. Only part of the ejected material sinks back onto the surface of Io. Water vapor is one of the prime movers of the eruptions of volcanoes on earth. But no water at all has been detected on Io. The Voyager data suggest that the driving forces of volcanic activity on Io must be gaseous sulfur dioxide and sulfur.

Volcanic activity on earth is also triggered by the heat generation of the radioactive elements present in the molten core. But the earth's mass is six times greater than Io's; it has therefore received larger quantities of radioactive substances from the primeval fog that filled the solar system. To keep a body the size of Io volcanically active for more than 4.5 billion years would require a superproportional presence of radioactive elements. Presumably the radioactive elements in the interior of Io still generate some heat. But this alone is surely not enough to explain the volcanic activity. It is assumed (but it remains no more than a theory) that the gravitational force of Jupiter and two of its moons, Europa and Ganymede, generates friction inside Io; this could provide the driving mechanism of the volcanoes. Voyager recorded temperatures of 20°C in the neighborhood of some of these volcanoes, whereas the normal temperature on the surface of Io is −138°C. These "hot spots" on Io could be lava lakes.

In addition to the active volcanoes on Io, Voyager also discovered hundreds of calderas, strikingly wide craters of extinct volcanoes enlarged to "cauldrons" through collapse or erosion.

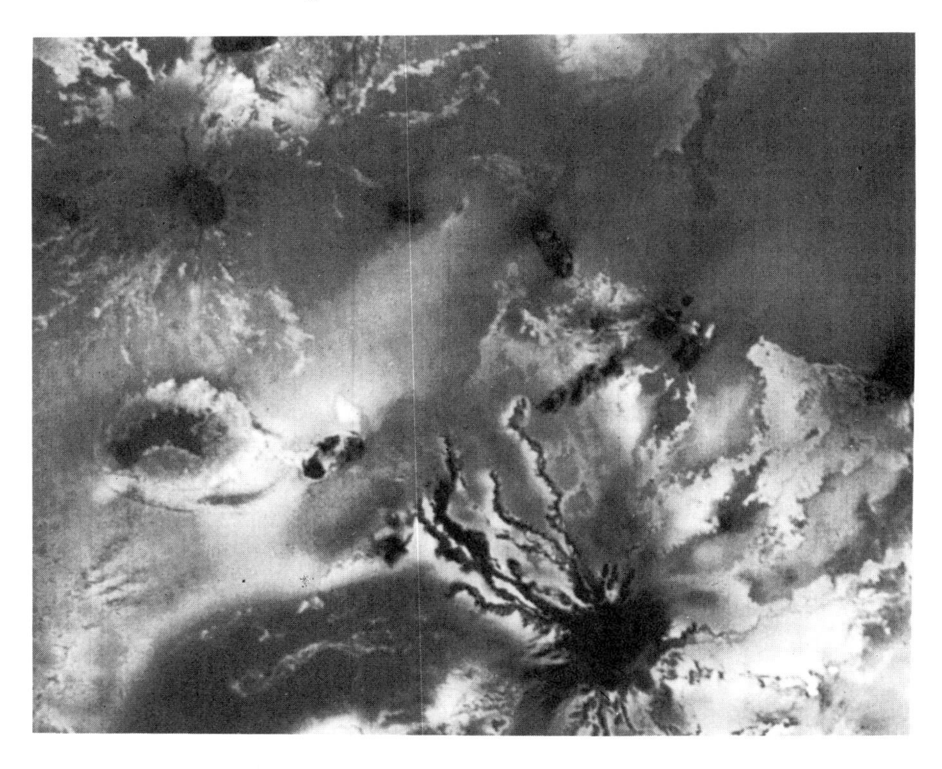

Io from a distance of 128,500km. The area shown has an edge length of about 100km. The diffuse colorations are reddish and orange and are probably due to deposits of sulfur compounds and other volcanic sublimates.

Some of these measure 200km across. Like the calderas on Mars, they are considerably larger than those on earth.

The quantities of sulfur Io ejects into space as a result of its volcanic activity are considerable. It has been calculated that this moon expels 3 million tons of sulfur, oxygen, and other substances per second. The satellite is entirely surrounded by a cloud of volcanic dust, which, orbiting Jupiter, together with Io, contains more than 4,500 particles per cc. By comparison, interplanetary space contains on average 1 particle per 10cc. The polar lights Voyager recorded on Jupiter can also be traced to the dust from Io's volcanoes.

Other interactions have been noted between Jupiter and Io. Not only does part of the volcanic ash leave the surface of Io for good; an electric current with an intensity of 5 million amps, discovered by Voyager 1, flows continuously between Io and Jupiter. It is a tubular region that joins Jupiter and Io magnetically. The electric current is probably generated by the rapid sweep of the Jovian magnetic field past Io, comparable to the generation of electric energy by the magnetic fields inside a current generator.

According to calculations by Eberhard Gruen and Gregor Morill (Max Planck Institute of Nuclear Physics), in collaboration with Torrence Johnson (Jet Propulsion Laboratory), particles smaller than 2/10,000mm—commonly found in volcanic ash—are electrically charged within a few seconds as soon as they reach altitudes of more than 100km above Io and have therefore left the thin atmosphere of this moon. This charge is due either to the ultraviolet light of the sun or, more likely, to bombardment by high-energy plasma particles in the Jovian radiation belts.

Because the magnetic field of Jupiter revolves at the same high speed as the planet on its axis, it sweeps past Io, which proceeds leisurely in its orbit around Jupiter, at great speed. This has the effect of a vacuum cleaner on the tiny, electrically charged particles of volcanic ash more than 100km above Io. They are forced along because, in proportion to their electric charge, their mass is too small for Io's gravitational force to retain them. Only particles between 0.1 and 0.01μm (1μm = 0.001mm) across suffer this fate. For larger ones Io's gravitational force exceeds the force of the electric charge; therefore, they return to the surface of Io. Smaller particles of ash, however, are completely pulverized by the particles in the radiation belts of Jupiter that strike them. Presumably these products of pulverization form the zone or train of plasma particles that Io constantly trails.

Dr. Gruen and his colleagues suspect that this zone also contains micromoons, remnants from the time the planetary system formed or of a moon that was broken up by the powerful tidal forces of Jupiter. Gruen and his colleagues call them "mother bodies" of the ring system, because, according to their theory, the ash particles hit them at speeds of 60km/sec and blast from them particles of up to 1,000 times their own size. The particles that form the ring around Jupiter, according to the calculations of scientists at the Max Planck Institute, must consist of particles of just this size. The scientists can point to some evidence of the accuracy of their theory. A fourteenth Jovian moon found, as already mentioned, beside moons 15 and 16 on pictures transmitted by Voyager, could be one of these "mother bodies." It orbits exactly in the narrow zone in which the scientists had predicted such objects. Its diameter of about 30km also makes it the size postulated by the theory. Dr. Gruen and his colleagues suspect the existence of numerous, altogether about 200, other such bodies.

Europa

The Jovian moon Europa is totally different from Io. It orbits the giant planet at a distance of 671,000km, or 9.4 Jupiter radii, once in 3.6 days. At a diameter of 3,130km it is about 10 percent smaller than our own moon. Europa is the smallest of the four Galilean satellites.

Europa displays a completely smooth surface. Dark, speckled, and uniformly bright terrain, broken up by many dark or bright narrow lines, can be distinguished. The lines are more than 1,000km long and 200 to 300km wide and could be faults or crevasses in the surface.

Only three volcanoes, each extinct and measuring about 20km across, were discovered on Europa. The major portion of this moon is covered with a layer of water ice, which once may have formed a liquid ocean flooding large parts of it at a time when Europa was still sufficiently heated by internal radioactivity to keep water in a liquid state. The surface of this ocean may have been covered with a thin sheet of ice. Today this ocean of ice surrounds the entire moon like an eggshell.

Ganymede

The third of the Galilean satellites, Ganymede, is the largest Jovian moon. Its diameter is 5,280km, more than 1.5 times that of the terrestrial moon. It orbits synchronously at a distance of 15 Jupiter radii or 1,1071,000km from the center of Jupiter.

Ganymede was photographed extensively by Voyager 1 on the side permanently facing Jupiter, and by Voyager 2 on the side facing away from it. We now have a good general picture of its topography. This moon is rich in surface detail from craters and linear structures that may have been produced by faults to large circular mountain ranges. The most striking object on Ganymede is a

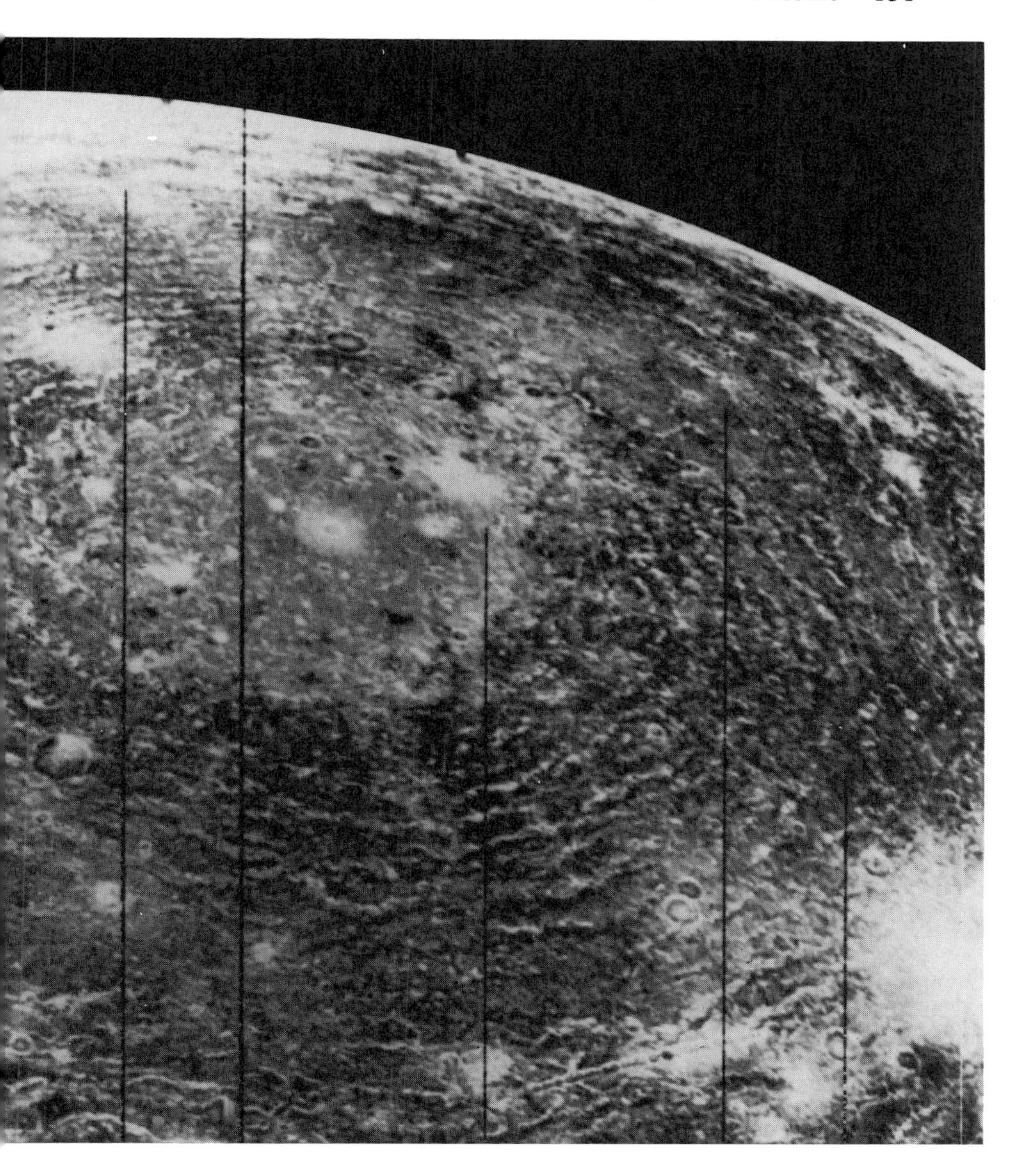

A photograph of Ganymede, Jupiter's largest moon, taken from a distance of 250,000km. The distance along the bottom edge measures about 1,000km. There are many impact craters and the surface is deeply furrowed. The furrows are often superimposed on other features, revealing their more recent origin. A large part of the surface is covered by what is thought to be ice.

circular basin with a diameter of 3,200km, resembling a mare on our own moon. This basin is crossed by narrowly-spaced white bands that look like furrows. Numerous secondary impact craters on Ganymede appear to be recent and intact. The aspect of Ganymede's surface is that of a ploughed field. The system of trenches is in parts rent asunder to widths of up to 50km, as if the original pattern had been destroyed by dislocations. This may indicate fault lines comparable to those on earth. Everything that can be seen on Ganymede points to gigantic changes that must at one time have occurred on this moon. Many structures are similar to the basins and craters on the inner planets of the solar system and on our moon.

The photographs taken of Ganymede by the Voyager probes show remnants of an enormous multiple circular mountain range that must have been caused by the impact of a very large body. Dark areas are covered with numerous impact craters with an estimated age of 4 billion years. In the lighter, furrowed areas, however, the craters number only about ten percent of those in the dark areas, indicating that the former are of more recent origin. There are no valleys and mountains on Ganymede. It would appear that the movement of large surface areas, comparable to the continental drift of the tectonic plates of our earth, ceased about 3 to 4 billion years ago. Spectrum analysis has revealed that Ganymede's surface consists of a mixture of rocks and ice. Its density suggests that half of it is made up of water ice.

Callisto

Callisto, the fourth Galilean satellite, has a diameter of 4,840km. It orbits Jupiter at a distance of 26.4 Jupiter radii, 1,884,000km, once every 16.7 days. It has the oldest surface of all the Galileian satellites, and at first glance it closely resembles our own moon and the planet Mercury. Callisto is completely saturated with meteoritic impact craters. Since the end of the violent meteorite bombardment about 4 billion years ago there have therefore been no major changes. Many annular structures are found on Callisto; these, too, indicate the impact of huge bodies billions of years ago. The annular basins are the equivalents of the maria on the terrestrial moon and on Mercury. But whereas the crust of our moon and of Mercury consists of silicates, that of Callisto is composed of ice and rocks. The density of this moon suggests that, like Ganymede, roughly half of it is made up of water ice. As on Ganymede, Callisto has no major mountains and valleys. Either the ice-rich crust of the moon cannot support such topographical features, or they were destroyed by the movement of the viscous ice.

If we consider the Galilean satellites together, we will notice that their mean density values decrease with increasing distance from Jupiter, similar to the mean densities of planets with reference to their distance from the sun. The innermost Galilean satellite, Io, has a mean density of 3.53g/cc, almost the same as that of Mars. Europa, the next innermost, has a density of 3.03. For Ganymede, the third, the value is 1.93; and for Callisto, the outermost, 1.79, which fairly closely approaches that of Jupiter itself, 1.3. Galileo's hypothesis that the satellite system of Jupiter is an image of the solar system is borne out also by its physical relations and points at cosmogonic, that is, evolutionary, connections.

The orbits of the other moons of Jupiter are much more distant from the planet. They are small bodies measuring between 7 and 80km across. None of them was investigated in detail by the Voyager probes.

SATURN—THE RING PLANET

Saturn can claim second place after Jupiter in the "hierarchy" of the planets: it is the second largest, the second most massive, the second fastest revolving planet. Its system of rings appears unique in the telescope (the discovery of rings around Uranus and Jupiter has made no difference because these ring systems are invisible from earth), and the brilliance of its surface details is second only to Jupiter's. Like Jupiter, Saturn shows bands of clouds, belts and zones, which are, however, far less distinct. Prominent objects of the type of the Great Red Spot have never been observed on Saturn.

Saturn is 9.946 astronomical units or 1,428 million kilometers away from the sun—almost twice as distant as Jupiter—and it does not appear very large even at the highest magnification in a telescope. The details of the ring system can, nevertheless, be distinguished and defined clearly even in amateur telescopes of today's quality.

This view of Saturn may be the best that can be obtained from earth. The picture was taken with the 3m reflecting telescope at the Lick Observatory on Mount Hamilton in California. The Cassini division is quite prominent, but we know that the ring system is much more complex.

Because the earth's and Saturn's positions in space change relative to each other, the earth is sometimes above, sometimes below the plane of the Saturn rings. As a result the aspect of the rings varies from year to year, making interpretation even more difficult. When the earth is precisely in the plane of the rings, they are completely invisible.

The Saturn ring has been known since the invention of the telescope, although Galileo, its discoverer, did not recognize it as a ring. "When I observe Saturn through a telescope of more than 30X magnification," he wrote to the ambassador of the Emperor of Austria on November 30, 1610, "it appears threefold; the largest star is in the center; the others are on a line east and west of it, appearing to touch it. They look to me like two attendants helping Old Saturn on his way, never straying from his side."

In 1656 the Dutch astronomer Christiaan Huygens recognized the real shape of the Saturn ring. Huygens also made the first rough sketches of Mars and discovered the Orion Nebula and Saturn's moon Titan. However, he saw the rings as a single, continuous object. A few years later it became clear that the ring actually consisted of several rings. Giovanni Cassini, a French astronomer, found a division in the Saturn ring, a gap that divided the "single" ring into two. Since then, this gap has been known as the Cassini division.

The Cassini division is more than 4,000km wide and separates the outer A ring from the bright, inner, B ring. A third ring—the crepe or C ring—was first seen in 1850. The crepe ring is inside the B ring and is seen only on very good photographs. This system of three rings was the "classical situation" of the Saturn rings. However, in September 1979 Pioneer 11 conducted a special investigation of the ring system and proved that the classical situation was no longer accurate .

It had been suspected that other rings—the D ring (inside crepe) and the E ring (outside A)—existed. Pioneer obtained very accurate geometrical measurements of the individual rings and found two more systems. One, a very wide outer ring, may be the same as the E ring. (If it is a new ring it will be called the G ring.) The second, also outside the A ring, was definitely identified as the F ring. Pioneer 11 did not confirm the existence of the D ring.

According to this new geometrical picture the innermost C ring begins at an altitude of 12,600km above the upper cloud limit of Saturn (to count from the center of Saturn add 60,000km to all data) and extends up to 32,000km. The B ring begins 32,000km above the cloud cover of Saturn and extends to 57,300km. It is followed by the about 3,600km-wide Cassini division; at its outer edge, at an altitude of 61,000km, the A ring begins. It extends to an altitude of 73,000km and is interrupted at 72,600km by the Encke division, no more than 870km wide, first seen in 1837 by Berlin astronomer Johann Franz Encke.

The F ring, discovered by Pioneer 11, is found at an altitude of 76,000 to 80,250km. The E ring (or G ring), discovered by the same space probe, reaches 270,000km into space. The roughly 3,700km-wide division between the A ring and F ring is called the Pioneer division in honor of its "discoverer."

Several hypotheses were established about the origin of these

On this photograph, taken by Voyager 1 on November 6, 1980, the Saturn ring system is seen to consist of ninety-five concentric structures. At the time of exposure the distance from Saturn was 8 million kilometers. The many details and structural features of the rings call for a new explanation; the system is much more complex than we had assumed.

rings. One thing is certain, they are not compact but consist of numerous pieces—similar to boulders. This led to the hypothesis that the ring is made of fragments of a moon or asteroid that approached the planet too closely and was therefore broken up into innumerable fragments; this theory was for a long time highly thought of as an explanation of how the Saturn rings formed. More recently, however, the view has been gaining ground that the objects circling Saturn might be residual material from the time when the planetary system evolved. The Pioneer investigations did confirm the old-established opinion that the Saturn ring is not compact, but consists of boulders of different sizes.

This idea of a Saturn ring made up of thousands of tiny moons can be traced back to Cassini, who in 1705 expressed the view that the Saturn ring was composed of "swarms of tiny satellites" too small to be seen discretely and orbiting Saturn at different speeds. It was not long before this view was confirmed.

Others, Sir William Herschel, for instance, claimed the opposite. Herschel not only thought the Saturn ring was compact, he believed that the Cassini division—he called it "the Black Belt"—was substantial; he would be prepared to accept it as a true gap in the ring only if a star could be observed in or through it. This is precisely what happened, although long after Herschel's death. In 1857, Cassini's theory that the ring was composed of a large number of moons was confirmed by a mathematical investigation conducted by British physicist James Clerk Maxwell. It was subsequently further borne out by spectroscopy. Additional research provided mathematical proof that the gaps, above all the Cassini division, were unstable zones in which no body could orbit for prolonged periods because of the gravitational forces of Saturn's moons.

The latest Pioneer 11 investigations have shed additional light on this complex of questions. At the time Pioneer 11 flew by Saturn the sun stood a little below the plane of the thin ring. Because of Saturn's motion around the sun, the position of the planet relative to the sun (as well as to the earth) changes; so, naturally, does the aspect of the Saturn ring. Here then the sun was slightly south of the plane of the rings, illuminating them from below, whereas Pioneer moved along a path that brought it toward the rings from above, that is, from the north. The sun's rays glanced off the ring particles from below, and to Pioneer 11 the rings appeared dark. This afforded the opportunity to determine the optical thickness of the rings, a measure of the transmission of light. From this we can derive data about the size of the individual particles that make up the ring.

Analysis showed that the rings are no more than about 5km thick and consist of bodies with diameters of up to 30m; presumably these bodies are composed of ice-covered silicate rock and perhaps ice-covered iron-nickel particles. The proportion of ice, at any rate, is considerable. The total mass of the Saturn rings is 570×10^{15} tons, about one millionth of the planet's mass.

This fly-by also yielded a large volume of information about Saturn's moons, charged particles, meteorites, dust particles, and the planet's magnetic field. In sum, it can be stated that this information largely conforms to that obtained by the Pioneer probes (and in even greater detail by the Voyager probes) to Jupiter. There are, for instance, interactions between the Saturn moons and the charged particles in the magnetic field of the planet, similar to the "vacuum cleaner effect" of the moons observed with Jupiter.

Saturn has the lowest density in the solar system, 0.69g/cc. At 1:10 it has the greatest oblateness of all planets. Its equatorial diameter, about 120,800km, is about 9.5 times that of the earth, but its mass is only 95 times that of the earth; this is due to its low density. The earth's density is 5.5; Jupiter's is 1.3g/cc. These density values suggest that Saturn's internal structure resembles Jupiter's: below an enormous envelope of gas, consisting of hydrogen, helium and slight amounts of ammonia and methane, at a temperature of $-150°C$ to $-170°C$ at its upper limit, the material should be in a liquid state with increasing pressure and temperature. Like Jupiter, then, Saturn does not have a solid surface. But the pressure in its center, because of the planet's smaller mass, should be lower than in Jupiter, at about 50 million bar at a temperature of 15,000° Kelvin. To round off the picture, Saturn's mean distance from the sun is 1,427 million km or about 9.5 astronomical units.

SATURN'S MOONS

Saturn probably has fifteen moons, although reports exist of up to twenty-two. Some cannot be definitely located, or their orbits are not known with certainty. Nine of the fifteen moons were discovered during the last century. The first to be discovered was Titan, by Huygens in 1655. Its diameter is 5,800km, which makes it larger than the planet Mercury. Its distance from Saturn is 1.22 million kilometers, or 10.38 Saturn radii.

The second largest of Saturn's moons, Japetus, was found by Cassini in 1671. Its diameter is 1,800km (about half the size of earth's moon), and it orbits at about 60 Saturn radii. The diameters of the other moons range between 1,450 and 170km.

With the exception of Titan all these moons are too small to retain an atmosphere. But an atmosphere was detected on Titan by Gerald Kuiper during observations in infrared light in 1944. It consists of methane, and its temperature is about $-180°C$. A number of speculations notwithstanding, the likelihood that this moon and its atmosphere could sustain life appears slight. At best, one could imagine early primitive life forms before the development of a terrestrial, oxygen-containing atmosphere. Moreover, Titan's atmosphere is probably not very dense. It cannot be very old, because, owing to high temperatures immediately after its origin the moon would have been unable to retain it. During that phase of the development of the planetary system the hot molecules moved rapidly enough to overcome Titan's gravitational force.

Another of Saturn's moons, Phoebe, is noteworthy because it follows a retrograde orbit—at a distance of 13 million kilometers and over a period of 550 days. No explanation has yet been found for the retrograde motion. It can, however, be assumed that Phoebe, like Jupiter moons VIII, IX, XI and XII (which also have retrograde orbits), is a planetoid captured by Saturn at a later stage or some other vagrant body of the solar system.

Like the perturbations between Jupiter, its moons, its radiation belts, and its magnetic field, numerous perturbations exist in the Saturnian system. But the most striking connection has been found between some of Saturn's moons and Saturn's rings. It has already been mentioned that the moons account for the existence of gaps between the rings. The outstanding example is the moon Mimas, with a diameter of 400km. Mimas orbits Saturn at a distance of 185,400km from the center of the planet and at a period of 22.6 hours. Within the Cassini division such a fragment would have an orbiting period of 11.3 hours—exactly half the time Mimas requires to orbit Saturn. But this means that the particles of the ring, were they present in this zone, would form a straight line with Saturn and Mimas every two revolutions. The resulting slight but constantly recurring attraction by Mimas of the particles that had been present in primeval times has over millions of years pushed them into different orbits, leaving the gap known as the Cassini division.

URANUS—THE GREEN-BLUE PLANET

The planet Uranus is more than twice as distant from the sun as Saturn—19.19 astronomical units or 2.869 billion kilometers. It moves at a velocity of 6.8km/sec and takes eighty-four years to complete its huge orbit around the sun. The Uranian day is 10 hours 49.5 minutes long. The polar axis (inclined in earth 23½°, in Mars

24°, in Jupiter 3°, and in Saturn 27° to the normal of their orbital planes) in Uranus has an inclination of 98°. This means that it is virtually in the planet's plane of rotation. Because the tilt of the axis from the vertical determines the seasons and the conditions of lighting from the sun on a planet, an extremely odd situation exists on Uranus. During the northern summer half year, which on this planet lasts forty-two terrestrial years, the North Pole points in the direction of the sun; during the winter half year the South Pole faces the sun. During the summer half year, then, the sun is almost constantly near the zenith of the North Pole; during the winter half it is above the South Pole, and the night lasts forty-two terrestrial years in the high northern latitudes. This must produce strange weather conditions, particularly because Uranus has a dense atmosphere consisting of huge amounts of methane, and smaller amounts of hydrogen, helium, and ammonia.

Owing to its great distance from the sun (and therefore from the earth) Uranus is a difficult planet to observe, always at the limit of visibility for the human eye. The planet was only discovered in 1781 by William Herschel, who thought at first that the faint object he observed through his telescope was a comet. He concluded from the circular orbit it described that he had found a new planet, one that orbited the sun beyond Saturn.

Although visual observation occasionally indicates surface markings on the planet, which in the telescope appears green-blue, it has so far been impossible to take photographs of any surface details. Perhaps we shall be more successful in January 1986 when Voyager 2 will fly by this faraway planet. At any rate, Uranus must be grouped with Jupiter, Saturn and Neptune as one of the outer giants of our solar system. Its density is low—1.21g/cc—its equatorial diameter is 50,800km, and its mass is 14.5 times that of the earth. The planet is enveloped by a deep, dense atomsphere of a composition similar to that of Jupiter's and Saturn's, differing only in the high proportion of methane. Uranus, too, may not have a proper solid surface although its core may be solid. The pressure in the center is estimated at 2 million bar; the temperature, at more than 4,000°C.

Two of the five known satellites of Uranus were discovered in 1787, two more in 1851, and the fifth in 1948. From the earth, the five moons of Uranus appear only as faint dots without any surface structure. Their diameters range between 600 and 1,800km; their distances from the center of Uranus, between 130,000 and 586,000km.

The discovery of a system of rings around Uranus was the outcome of indirect observations in 1977 and is becoming more and more mysterious. It began with an attempt in March 1977 to obtain better information about the diameter of Uranus during a period of occultation—a time when Uranus would pass, and thus obscure, a faint star. Half an hour before the event, the star disappeared five times for short periods. This was repeated half an hour after the occultation. The astronomers, who carried out their observations from a NASA research plane flying across the Indian Ocean, concluded that Uranus must be surrounded by five rings too weak to be visible in a telescope but observable when they attenuated the light from a star in the background.

During further star occultations by Uranus in December 1977 and in 1978 the observations were repeated from airplanes and balloons and as a result the number of suspected rings increased to nine. But all these rings are relatively close to the planet, very tenuous, and rather inconspicuous. Seven of them are only 5km wide, one measures 15km, and another perhaps 70 to 100km. Their thickness, too, is no more than a few kilometers, the width of the entire nine-ring system, which starts at an altitude of 16,580km above the cloud cover of Uranus, is only 9,720km.

It was originally assumed that these rings—only 1/1,000th as wide as those of Saturn—consist of snow or ice and that their individual components have diameters of 1 to 2km. But the low albedo, the reflectance of the rings or of their components, contradicts the snow or ice theory. The maximum albedo is only 3 percent. Only the blackest coal dust has such a low albedo, and so far we know of no substance in space of such low reflecting power. Up to now there is only speculation about how such a ring system can be stable. The narrowness of the individual rings should have long resulted in their widening and attenuation owing to mutual perturbation of their components. The American astronomer van Flandern therefore suspects that the Uranian rings do not consist of solid lumps of matter but of gases continuously emitted by satellites and by the latters' gravitational force held in satellite orbits for some time. The observed attenuation of the light from stars behind them, which led to the discovery of the rings, would therefore be due only to the diffraction of the starlight as it passes through these gases, rather than to an occultation by denser matter. In 1986 Voyager 2 may shed more light on these problems, if it is still capable of collecting and transmitting data.

NEPTUNE—THE OUTERMOST GIANT PLANET

We must add another 10 astronomical units beyond Uranus to reach Neptune, the second outermost planet of the solar system; 30.06 astronomical units, or 4.497 billion kilometers, separate it from the sun. This remote planet requires 164.8 terrestrial years to complete its almost circular orbit. Its discovery is a much admired example of the efficiency of theoretical astronomy.

Neptune was not found in the sky with a telescope by mere accident but "with a nib of a pen." From irregularities in the orbit of Uranus observed at the beginning of the nineteenth century, two astronomers had deduced perturbation by another, still unknown, planet. John Couch Adams in Britain and Urbain Leverrier in France claimed that the deviations of Uranus from its theoretically calculated orbit were due to the gravitational force of another planet. From the suspected perturbations caused by this unknown planet the two astronomers computed its orbit. On September 23, 1846, the assistant at the Berlin Observatory, Johann Gottfried Galle, received a letter from Leverrier indicating the position the unknown planet would have to occupy in the sky on the basis of his calculations. On the same evening Galle discovered the new planet, less than one arc degree away from the stated position. This was one of the greatest triumphs of theoretical astronomy, a discovery that we owe to classical mechanics and to Newton's Laws of Gravity, methods that enable us to predict the existence of undiscovered planets.

The resemblance between Uranus and Neptune is extraordinary. They are of almost identical size (Neptune's equatorial diameter is 49,000km; Uranus's is 50,800km), have comparable densities (Neptune 1.7, Uranus 1.21g/cc), and masses (Neptune 17.2, Uranus 14.5 earth masses). It takes Neptune 15 hours 49.5 minutes to revolve once on its axis, which at 28°48′ is slightly more inclined to the vertical of the orbital plane than the earth's.

Through a telescope, traces of detail, without doubt not of Neptune's surface, but of cloud formations in the atmosphere, can be observed. Like that of Uranus, the atmosphere contains a high proportion of methane as well as hydrogen and helium. Neptune also appears green, due to the absorption of yellow and red light by the atmospheric methane. At the upper limit of Neptune's atmosphere the temperature may range between $-130°$ and $-200°$. Neptune has perhaps a rocky core with a diameter of about 16,000km, surrounded by a layer of ice about 8,000km thick.

Neptune is accompanied by two moons; the inner, Triton, is the largest moon of the entire solar system. It has a diameter of 6,000km and completes its retrograde orbit in 5.9 days at a distance of 353,000km.

Neptune's second moon, Nereid, moves at an average distance of 5.6 million kilometers from its planet. Its diameter is only 500km; its orbit is very eccentric, and takes 359.4 days. As a result of the great eccentricity, Nereid's distance from Neptune varies between 1.33 million and 9.76 million kilometers. The orbital data of the two satellites are unusual enough to suggest an unusual history of these bodies.

PLUTO—RESULT OF A COSMIC CATASTROPHE?

The history of Pluto's discovery is almost a copy of that of Neptune—but only "almost." Even after the discovery of Neptune in 1846 deviations were noted from the theoretical motions of Uranus and Neptune. They made Percival Lowell compute the orbit of a hypothetical "Transneptune." On the basis of his calculations Lowell began the search for this planet. Its successful conclusion, however, was posthumous; in February 1930 Clyde Tombaugh at the Lowell Observatory found a tiny object of the 14th magnitude on several photographs, moving slowly and thereby revealing its planetary nature. Tombaugh was able to confirm that it was the suspected outermost planet of the solar system, not far from the calculated spot.

But very soon it became obvious that Pluto did not at all fit into the picture we have of the outer Giant Planets. Pluto has a diameter of about 3,000km, which makes it smaller than Mercury, smaller even than our own moon. It follows the most eccentric orbit of all the planets. Its mean distance from the sun is 39.53 astronomical units, or 5.95 billion kilometers. But its minimum distance from the sun is 4.425 billion kilometers and its maximum distance is 7.375 billion kilometers. This means that when Pluto is nearest the sun it is nearer to it than Neptune; Pluto's and Neptune's orbits intersect. At 17.8° to the ecliptic (the apparent path of the sun in the sky) Pluto's orbital inclination is far greater than that of any other planet of our solar system. Its density is very low (1.0, the density of water). The planet has a small moon discovered in 1978 and given the name of Charon. Charon has a diameter of 850km and orbits Pluto in the same period as the planet revolves on its axis, 6 days, 9 hours, 17 minutes. We can deduce from this strange agreement between the period of Pluto's revolution and of the orbit of its companion that Pluto is egg-shaped, not spherical.

The special features of the orbits of Triton and Nereid around

The arrows on this photograph indicate the location of Pluto, which looks here like a tiny star. The picture was taken on March 4, 1930, a few weeks after Clyde W. Tombaugh (at the Lowell Observatory) discovered Pluto.

Neptune and the unusual orbit of Pluto around the sun suggested to some astronomers that Pluto could be a former moon of Neptune that had been ejected from its original orbit. Harrington and van Flandern made this hypothesis the basis of computer calculations of a model in 1978, which assumes that Triton, Nereid, and Pluto were originally satellites of Neptune that moved in normal orbits. But the two astronomers suggest that a long time ago a large body passing Neptune perturbed the orbits of the three moons. Their computations showed that a close encounter of a massive body with Neptune would result in orbits described by the three bodies today. They also showed that Pluto would break up into two uneven parts, into Pluto itself and into its moon Charon; this fragment would move around Pluto in a synchronous orbit precisely as it does.

EARTH—THE WATER PLANET

In our ramble through our planetary system we have met many celestial bodies of the most diverse properties—cosmic hells, cosmic ice deserts, cosmic witches' cauldrons, and cosmic emptiness. We have found states, conditions, and forms of matter which on our own planet are quite often unimaginable. We have learned that our earth is unique—at least in our own solar system. The most fitting description some extraterrestrial intelligent being could find for our planet would be "the water planet." Nowhere else in our solar system do we find a body of such strange qualities. More than 70 percent of the earth's surface is covered by what may be the oddest liquid in the universe, a compound of two atoms of hydrogen and one atom of oxygen, $H_2 0$—water.

It seems that Mars has already made the acquaintance of water; it is possible that Venus will contain water at some time in the future. At present, however, and certainly for billions of years in the past and to come, the earth is the only "water planet." No less special is the earth's atmosphere and its high content of oxygen, essential to advanced forms of life, a deadly poison to many primitive ones. The earth is a wet planet everywhere, not only where the oceans cover it. Compared with the vast plains of Mars, the stony wastes of the moon, or the craters of Mercury even the Sahara is a soaked sponge. The earth is subject to continuous change; clouds form here, then somewhere else, there are gales and blizzards, the land-

This picture of earth was taken by Apollo 9 from an altitude of about 350km on March 11, 1969. It shows a cyclone less than 2,000km north of Hawaii. From space, life on earth could be demonstrated only by circumstantial evidence. The earth's atmosphere looks similar to that of Venus, Jupiter, and Saturn.

scape bursts out in shimmering green contrasting with the luminous blue of the oceans.

The earth's atmosphere is 21 percent oxygen, 78 percent nitrogen, and only 1 percent miscellaneous gaseous elements. This is unique. When we look at the atmosphere of Mars and Venus we find that it consists mainly of carbon dioxide (97 and 95 percent respectively), in large doses a lethal substance for creatures living on earth today. The earth's atmosphere contains no more than 0.03 percent. Oxygen is present in traces in the Venusian and in the Martian atmosphere. As far as we know the high oxygen content of the earth's atmosphere alone is an indication of the presence of life on earth. It is the product of unceasing biological activity on earth and has been evolved during more than 3 billion years of life processes. Photosynthesis, the conversion by plants of inorganic into organic substances with the aid of sunlight, produces enormous amounts of oxygen. This is a reaction which at the same time uses up carbon dioxide. At present we know of no process that produces large quantities of free oxygen in a planetary atmosphere through other than biological reactions. To visitors from space the oxygen content of the earth's atmosphere would be clear circumstantial evidence of the presence of life on this planet.

Is this life an accident or the result of a purposeful development? We do not know. What we do know is that it has existed on this planet for a considerable time, that life today is the outcome of a long process of evolution on this planet. But why only here? Why not on the other planets and moons of our solar system? Has life really had such limiting conditions imposed on it that it was able to unfold only on earth? In view of what we can see in our own solar system this question may appear compelling. Yet we still know far too little about the history of our world (and much less about its future) to answer this question.

The questions: How did life begin? How widespread it is in the universe? For what period of time it is able to last on a given planet? lead to questions of the origin and evolution of our own planetary system and indeed of the history of the universe as a whole.

A SOLAR SYSTEM IS BORN

Five billion years ago our corner of the universe was dark and bitterly cold. No sun, no planets existed; almost empty dark space stretched for billions and trillions of kilometers in every direction. This emptiness was occasionally broken by a few stray atoms embedded in a universe in which primeval suns glowed at unimaginable distances. These suns were so remote that they looked like dots and were no brighter than the present stars; there were merely fewer of them. While today there are 3×10^{18} atoms in one cubic centimeter of air, at that stage these atoms would be scattered over a space of 2,500 cubic kilometers.

Matter, then, was extremely rare. Besides the atoms—75 percent of which were hydrogen, and together with helium forming 95 percent of all existing atoms—there were only a few percent of dust particles measuring about 0.001mm across, and only about 100 of them in every cubic kilometer. With the atoms of other elements they formed 5 percent per unit weight of that huge cosmic cloud of gas and dust that, because of its negligible particle density, nobody could have detected. The dust granules consisted of silicon, aluminum, iron, oxygen, carbon, nitrogen, and occasionally organic molecules, that is, of practically all the substances making up our world today, substances essential to living creatures.

But there was as yet no trace of the—let us call it divine—spirit that created from all this a radiant sun, planets, moons and, ultimately, living organisms 5 billion years ago. The gas pressure inside this primeval cloud was balanced by the centripetal mutual gravitational force of the particles; there was a cosmic equilibrium. The starting signal for the creation of a sun, indeed of an entire solar system, the triggering agency capable of upsetting the equilibrium of the cloud and of allowing celestial bodies to condense from the masses of gas and dust had yet to make its entry upon the scene.

However, the cosmic spark was ignited; the cloud giving birth to our planetary system was set in motion. Nobody knows exactly how this happened, but there are two theories. The first theory posits that an arm of our galactic system, the spiral nebula to which we belong (a gigantic cloud of 100 billion suns in the shape of a spiral) swept through our region of the universe and caused a slight condensation of the molecules and atoms it encountered. The second theory suggests that the shock wave of a faraway stellar explosion—perhaps of a supernova—assisted in the birth of our solar system. There is support from astronomical observation for both theories, and it is therefore not yet possible to decide which of these two processes triggered the formation of our sun, planets, and moons.

The shock wave created by the spiral arm or by the supernova condensed our cosmic protocloud about 100 times; the 100 dust particles previously found in one cubic kilometer closed up to become 10,000. The cloud, at first receiving little heat from the distant stars, became opaque; because of its greater density its temperature, already near absolute zero, dropped even further. As a result, gas pressure also dropped. The gravitational force gained the upper hand; this produced a further condensation of the cloud. We know of numerous such cold, dark, contracting clouds of interstellar gas and dust in space. Our cloud began to move; vortices formed; matter condensed and broke up into individual rotating parts, and one of these cloud fragments became our solar system.

All this must be seen in terms of an extremely slow, gradual process. Our fragment of gas and dust took shape, became our pristine solar nebula. It had a diameter of about 10 billion kilometers, extending roughly to the present orbit of Neptune. At 200 million kilometers its substance was about twice that of today. The gravitational force between the atoms, molecules, and dust grains continued to have the upper hand over gas pressure; especially toward the center of the cloud, matter condensed sufficiently to produce a local rise in temperature. The temperature gradient from the interior of our protonebula—where matter was very hot—to the edge—where it was very cold—created the basis for the different properties of the future planets.

The entire process passed through several phases. During the first 50 million years after the triggering shock wave, the solar nebula formed and was followed by a further concentration of matter in the center—the protosun. In the hot plasma magnetic fields were set up; these distributed the increasing angular momentum the protosun acquired from rotation and contraction to the marginal regions of the solar nebula and established the preconditions for the birth of the planets. The key to the different natures of the planets already existed at this stage. It was provided by the difference in the temperatures of the protonebula; high in the center, lower toward the edges.

A result of this temperature distribution and of the angular momentum of our protosun throughout the entire solar system (a process that took only a few thousand years) was a tidy separation of substances according to their melting and boiling points. During the formation of our protosun the temperature in the center of our planetary system was several thousand degrees centigrade so most substances evaporated. In the outer regions of the protonebula, however, temperatures were no higher than −150 to −180°C. Here the dust grains, consisting of silicon and its compounds, were as unaffected by melting processes as aluminum, magnesium, and comparable chemical elements and their compounds of high melting point. They became coated instead with layers of ice, frozen methane, ammonia, and similar substances.

When the protosun was formed and gradually began to cool, those substances in its surroundings that had the highest melting points, such as ores and oxides of metals—solidified. But the farther out these nuclei of condensation were, the greater the proportions of volatile substances they contained. In these various compositions they filled the entire volume of the solar protonebula.

Numerous collisions took place between the individual fluffy grains, which as a result loosely "baked" together. In the course of time this produced lumps with diameters ranging from a few millimeters to several centimeters. During some hundreds of thousands of years the majority of them collected in the plane of the protonebula under the influence of gravitation. They occupied a huge area, with the protosun in the center. Gravitation gradually produced regions of condensation, where many small lumps congregated; other zones became emptier, with the grains migrating to more densely populated areas. There were numerous collisions; step by step the small lumps became larger, developed into objects a kilometer wide, into planetesimals. They formed the bodies of condensation for the protoplanets proper. The chemical composition of the planetesimals, however, already differed widely. Again, it simply depended on the location of these structures. In regions near the sun they consisted mainly of elements and compounds that do not readily melt; in more distant zones such volatile substances as frozen methane and frozen ammonia predominated, exactly as in the first dust grains at the beginning of condensation into millimeter-sized objects.

During further millions of years, as a result of gravitation, the planetesimals congregated and produced much larger bodies—the protoplanets. Four of these protoplanets formed in the inner solar system, four in the outer. The four inner protoplanets later became Mercury, Venus, Earth, and Mars; the four outer ones became Jupiter, Saturn, Uranus, and Neptune. In the wide gap between the orbits of Mars and Jupiter there were hundreds of thousands of planetesimals or early protoplanets. Here, too, a planet probably attempted to form, but for some reason failed to do so.

The protoplanets were the products of the conglomeration of planetesimals owing to the general gravitational force and ultimately developed into today's planets. Accordingly, our solar system must have been literally teeming with planetesimals for such large bodies to have evolved through attraction and collision. Initially these protoplanets were "cool"; they were not heated internally, and their temperatures corresponded with those of the protonebula around them. In the neighborhood of the sun this temperature was 1,650°C; in the vicinity of the earth it may have been about 300°C; and further out, in the region of Jupiter and Saturn, lower than −100°C.

But the more material accumulated while the protoplanets developed, the larger the conglomerations became, the more the temper-

The Asteroid Belt

The old assumption that these asteroids or planetoids are debris of a larger planet that exploded is for various reasons no longer tenable. At present about 2,200 asteroids are specifically known to us; their orbits have been computed so that they can be located again. About 5,000 asteroids have been discovered, but in reality there may be 50,000 or even 100,000 of them. They orbit the sun mainly between the orbits of Mars and Jupiter, but there are exceptions. The orbits of some asteroids almost approach that of the earth; those of others extend beyond the orbit of Jupiter. The largest asteroids have diameters of about 500 to 1,000km, but most of them measure between 100 to 30 or so kilometers. There is probably a continuous transition to the meteorites and therefore objects of only a few hundred, or even a few meters diameter. Only 230 of the known asteroids have diameters larger than 100km. The total mass of the asteroids is no more than about 1/10,000th that of the earth. This very figure militates against the assumption that they are debris of a single former planet. Where then are the additional fragments that must have gone into making up a planet of reasonable size?

There are much stronger reasons to regard the asteroids as remnants of protomatter that, although condensed into planetesimals from the original solar nebula, did not manage to form a planet. The asteroid belt would therefore be a structure comparable with the Saturn rings. It is remarkable in this context that there are gaps in the asteroid belt corresponding to the divisions in Saturn's rings. These are resonance orbits of the planet Jupiter; that is, orbits in which the orbital period of an asteroid would be a simple fraction of that of Jupiter, and where as a result all objects that had previously been there would have been expelled owing to orbital perturbations of Jupiter.

ature of these new worlds rose owing to the radioactive decay of the elements in their interior. This radioactive heating, which is still continuing in some planets, led to the melting of the individual bodies from inside. Because of the gravitational force the heavy elements sank to the core of these planets. This especially affected iron, nickel, and substances of similar specific gravities. On the other hand, the lighter elements were driven outward from the core into the upper layers of the protoplanets. This is how the familiar chemically differentiated arrangements of substances in the interior of planets, with cores of iron surrounded by lighter layers of rock, originated.

In Mercury and Mars this process developed even further. Because of their weak gravitational forces these planets were no longer capable of retaining highly volatile substances such as hydrogen and helium at the high temperatures caused by the melting; the rocks melted, but these elements evaporated. Only Venus and Earth proved massive enough to retain at least minor atmospheres, mainly 10 to 15km above their surface. These atmospheres consisted largely of carbon dioxide—even Mars was able to retain a quite negligible, extremely thin residual layer of this compound. The unique development of plant life on our planet must be held responsible for the completely different chemical composition of the earth's atmosphere.

Even before planetesimals and protoplanets were formed a separation of the elements, as already mentioned, was already taking place as a function of temperature in the protonebula: the higher the temperature, the higher the proportion of elements with high melting points. Iron, nickel, and various metallic oxides were the first to condense in the vicinity of the protosun, followed by the somewhat cooler rocks. Farther out in the solar system—where the protoplanets of Jupiter, Saturn, Uranus, and Neptune formed and where, compared with the inner solar system, it was bitterly cold—even substances as volatile as methane, ammonia, and ice condensed. In addition, the huge protoplanets, with their powerful gravitational force, "swept" the space around them "clean" and incorporated all gases down to helium and hydrogen. This is why the giant planets have enormous atmospheres of methane and ammonia. The largest planets, Jupiter and Saturn, also added the light gases helium and hydrogen to their atmosphere. The somewhat smaller planets Uranus and Neptune, although still huge compared with Earth or Mars, were not able to capture as many light elements as Jupiter and Saturn; they had to be content with the thick atmosphere of methane and ammonia.

During the formation of the protoplanets a process of creation took place in the center of the solar system without which the evolution of the future planets would have had neither rhyme nor reason and which was the precondition of the development of life on earth. Because of the gravitational force the protosun progressively collapsed and became hotter and hotter until finally a pressure of millions and billions of atmospheres was set up in its core. It triggered an elementary fire, the fusion of hydrogen nuclei into helium nuclei and the resultant liberation of energy. The core of the sun assumed a temperature of 15 million degrees Centigrade, and an unimaginable quantity of light and heat began to flood the solar system. This took place about 4.5 billion years ago and was the hour of birth of our sun.

The region in which the solar system was formed was still teeming with particles of the protonebula; after all, 95 percent of its matter consisted of hydrogen and helium. But now, after the thermonuclear firing of the sun, an interplanetary hurricane began to rage. With inconceivable force the sun ejected great quantities of hydrogen and helium nuclei. For millions of years this stream of particles swept far out beyond the solar system and took with it remnants of the protonebula and everything else that blocked its path. It overcame the original atmospheres of the planets and blew a large part of them away. Only the planets, the moons (which had similarly formed from the protonebula), the asteroids, and other fragments of matter—comets, large meteorites—were able to withstand this primeval storm.

As the inferno abated the creation of the planetary system was largely accomplished. Whatever happened afterward was no more than "cosmetic," a number of isolated reactions without fundamental significance.

Our Sun

THE SUN—PRIME MOVER OF OUR PLANETARY SYSTEM

The sun, the center of the planetary system, is the prime source of all life on earth. Without the sun there would be no life on our home planet; there would be no solar system. The gravitational force of the sun keeps the planets in their elliptical orbits; provides them with light and heat energy and thereby causes movements in the planets' atmospheres; creates climates and the weather; and triggers storms and winds in envelopes of gas which, without solar heat, would be deep-frozen of solids carbon dioxide, methane, ammonia.

All the forms of chemical energy we use on earth are derived from solar energy; the lumber we fell and burn, like the grass, the plants and the trees which grow and flourish in its warmth and with their energy convert inorganic substances into organic matter—all owe their existence to the sun. During the Tertiary period millions of years ago oil was produced by a process of decay of tiny animals and plants that owed their life to the sun and were embedded in sand and mud. Coal, too, has its origin in plants that populated the surface of the earth.

We must also thank the sun for our wind and water power. It is the sun that causes temperature and pressure differences in the atmosphere, and thus the winds that try to compensate for these differences. It is the sun that makes water evaporate and lets it return to rivers and oceans in the form of rain, where it drives the turbines of our hydroelectric power stations. There is only one form of energy that does not come directly from the sun—although it originates in the same forces from which the energy of our sun is derived—nuclear energy.

For many years astrophysicists puzzled over the sun's ability to radiate, for billions of years and at absolute constancy, a quantity of energy that supplies every square meter of the earth—with vertically incident rays and without the energy-absorbing effect of the atmosphere at 1.36kW or, to express it in the previously used measuring unit, 1.95 calories per square centimeter/minute. In practical terms this means that 1cc of water will be heated at a rate of almost 2°C per minute by the solar radiation incident on an area of 1 square centimeter. The entire solar energy that reaches the earth's surface amounts to about 170 trillion kW (in figures, 170,000,000,000,000, abbreviated, 1.7×10^{14} kW). Exponential notation is used when very large or very small numbers are to be expressed; for example, 10^{14} (ten to the power of fourteen) is a 1 followed by 14 zeros. Hence, $10^2 = 100$, $10^3 = 1{,}000$, $10^4 = 10{,}000$, $10^6 = 1{,}000{,}000$ and so on. Accordingly, $10^1 = 10$, $10^0 = 1$, $10^{-1} = 0.1$, $10^{-2} = 0.01$. This notation is also used in equations; for example $5 \times 10^6 = 5$ million; $1.7 \times 10^{14} = 170$ trillion. Compared with the 170 trillion kW of the sun, the energy consumption of the whole of mankind on household, industry, transportation, lighting, heating, and mechanical power is truly negligible.

But even the 170 trillion kW is only a tiny fraction of the energy that the sun radiates into space in all directions every second. This total of emitted solar energy is also called the absolute luminosity of the sun. It is 4×10^{23}kW, or 400 sextillions. Because of its distance from the sun, the earth receives no more than two trillionths of it. Astronomic magnitudes and distances often have the disadvantage of begging all human imagination—simply because they are so far beyond our range of daily experiences. Only after prolonged contact with them do we vaguely appreciate their true meaning.

These two trillionths of the sun's luminosity arrive at the upper boundary of the earth's atmosphere, where about 30 percent of it is absorbed. A large part of these 30 percent consists of the shortwave ultraviolet radiation of the sun, x-rays, and various other types of radiation invisible to the human eye. The quantity of absorbed solar energy ensures the circulation of the troposphere, the atmospheric movements at altitudes of 10 to 17km that produce our weather.

The tower telescope at Mount Wilson Observatory (near Los Angeles, California) is one of the solar telescopes that have made history in solar research. A coelostat—a mirror system in the dome—directs the sun's rays into the observation room 50m below. Here, cameras and spectrographs allow detailed observation of solar phenomena.

As already mentioned, the distance between the earth and the sun is 149.6 million kilometers. This is 375,000 times the equatorial circumference of the earth and about 390 times the distance between the earth and the moon. If we were able to cover the distance from the earth to the moon in an express train traveling 100kmph, hour after hour, day after day, we would travel for 160 days. To cover the roughly 150 million kilometers that separate us from the sun we would have to spend 170 years on our express train. Even in a jet liner flying at 1,000kmph it would take us 17 years to reach the sun.

The light ray travels at a velocity of about 300,000km/sec—the highest velocity possible according to Einstein's Relativity theory and a speed that cannot be reached by any material object—and takes 8.33 minutes to travel from the sun to the earth. We use the time required by light to cover a certain distance also as a measure of distance—the light year; for example, or the light minute—and say that the sun is 8.33 light minutes away from us. And in spite of this huge distance the sun still appears to us as a disk the size of the full moon in the sky—its apparent angular diameter is on average not quite 32 arc minutes or half an arc degree—indicating that the true diameter of the sun must be huge. It measures 1,392,000km, 109 times that of the earth. We would need 1,300,000 spheres the size of the earth to fill a sphere the size of the sun. These figures bring home the reality of the gigantic dimensions of all the reactions we can observe on the sun. This sun is a huge ball of gas consisting mainly of hydrogen and helium. The sun has no solid surface; owing to the temperatures prevailing on and in the sun all chemical elements are in a gaseous state. Each square centimeter of the sun's surface continuously emits 6.3kW of radiant energy; the radiant temperature of the sun is 5,770° Kelvin. The so-called absolute temperature is measured in degrees Kelvin, with absolute zero at −273.16°. For the conversion of degrees absolute into centigrade, 273°C must be subtracted; conversely, 273 must be added to the centigrade value to obtain the absolute value.

It is obvious that the question of the type of energy source that is capable of maintaining the sun at such a high temperature is closely related with the time during which the sun has already emitted such radiation. Many considerations and investigations concern the question whether the intensity of the solar radiation fluctuates in the course of time and it if does, to what extent. All the results emphasize that such fluctuations of the solar constant, if they exist at all, are negligible. This has been proved both by measurements over the last hundred years and by geological evidence from the remote past of the earth. Fluctuations of the solar constant would, after all, result in climatic changes on earth, which in turn would have far-reaching effects on flora and fauna; this could not fail to be reflected by historical evidence. But there is no indication of really severe fluctuations of the amount of energy supplied by the sun to earth during the last few thousand million years. In fact, we have proof that solar irradiation of the earth 3.5 billion years ago was practically the same as today. Geological finds have revealed, for instance, the existence at that time in South Africa of single-celled organisms whose structure was about as complex as that of the modern blue algae. To enable them to develop, a solar irradiation of about present-day intensity was necessary. Climatological investigations also confirm the relative constancy of the solar irradiation of the earth. The negligible irregular fluctuations that have been observed are below 1 percent and therefore are of no interest in this context.

THE ENERGY SOURCE OF THE SUN

It is established that for about 5 billion years the sun has radiated the same amount of energy. Where can we look for the source of this energy?

Peculiar hypotheses were developed to explain the thermal economy of the sun. Calculations had shown comparatively quickly that a chemical "burning" of the sun, like a huge coal fire, would provide no better explanation for the long-term energy emission of the sun than all the other chemical reactions. Even if the sun had consisted of the purest coal, its life expectancy would have been no more than 5,000 years, and it would have burned out long ago. A number of other exotic hypotheses—such as a solar "furnace" constantly stoked by the meteorites and comets that plunge into it—are equally unsatisfactory.

Another theory, that of the German physicist Hermann von Helmholtz, can be taken more seriously. Helmholtz suggested that the sun derived its energy from a continuous process of contraction, becoming smaller all the time. But here, too, calculation shows that this gravitational energy would at best be sufficient to maintain the sun for no longer than 10 million years.

In both nuclear fission and nuclear fusion part of the matter involved in the transformation seemingly "disappears." If we place a heavy atomic nucleus—of uranium or of plutonium, for instance—on a balance before it undergoes fission and weigh it again after, placing all the fragments produced and all the other products liberated on the balance, the total mass of the parts would not add up to that of the original atom. Conversely, four hydrogen nuclei (which during fusion would be transformed into one helium nucleus), weigh a little more than the helium nucleus. These differences in mass are converted into energy according to the famous equation, established by Einstein, $E = mc^2$, where E = energy, m = the transformed mass, c = the velocity of light.

Only the combination of our knowledge of nuclear physics with that of astronomy has helped us to understand the reactions inside the sun. This knowledge, far from consisting of hypothetical assumptions, is therefore based on theory and practice and confirmed by experiment and observation preceded by wide-ranging considerations and detailed calculations. If we can now claim with assurance that the energy source of our sun is the nuclear fusion of hydrogen into helium, this is the result of detailed information about the possible nuclear transformations and the comparison of this information with the actual conditions prevailing on and in our sun.

The reactions between the atomic nuclei are not random, but the consequence of clearly defined predictable states and preconditions that allow us to compute models of stars and to see what happens to them in given circumstances. This is precisely what Hans Bethe in the United States and Carl Friedrich von Weizsaecker in Germany did in 1938. To begin with they discovered a complex sequence of transformations, in the course of which during six nuclear transformations a helium atom is built up from four hydrogen atoms with the help of a carbon atom. The latter emerges from this process unaffected; it acts as a catalyst. A year later Bethe listed all the nuclear transformations that could take place in the sun. In doing this he found a second way in which hydrogen nuclei can fuse into helium nuclei and in 1967 was awarded the Nobel Prize for this discovery. For, at least in our sun, this second way is probably the more important one.

Let us examine these two sequences in greater detail. The cycle that is more important in our own sun is called the H + H reaction (H = hydrogen), the fusion of two hydrogen atoms to form a nucleus of the heavy hydrogen deuterium, called H^2, emitting one positron and one neutron. The formula is written as follows:

$$H_1^1 + H_1^1 \qquad H_1^2 + e^+ + \nu$$

The subscripts are the atomic numbers and indicate the number of positive particles—that is, the protons—in the atomic nuclei, thereby defining the chemical nature of each element. The superscripts are the nuclear numbers and indicate the total of all the elementary particles forming the atomic nucleus (protons and neutrons). The hydrogen atom H, for instance, has only a single proton. H_1^2 indicates an isotope (a different version) of hydrogen, in which the atomic nucleus consists of a proton and a neutron. Here this isotope is produced by the emission of a positron—a positively charged electron—by the two H nuclei. The symbol ν is a clear indication of the presence of a neutrino, a zero-mass, zero-charge particle also produced during the reaction.

In the course of his investigation Bethe calculated the probabilities of the various possible nuclear reactions. They are the function of some atomic constants (the previously-mentioned nuclear-physics aspect of the problem) and of the conditions inside the sun (its astronomical and astrophysical aspect). On average one H + H reaction will occur once every 7 billion years. Nevertheless, it is notable because of the enormous number of hydrogen atoms that make up the sun. Many reactions can occur simultaneously, beginning and ending at different times; so in fact the reaction takes place frequently. The initial conditions on which Bethe based this model assumed a temperature of 13,000,000°C in the interior of the sun and a density of matter of 100g/cc. The calculation also shows that the probability of this reaction increases greatly at even higher temperatures and that at an assumed temperature of 100,000,000°C (which does not exist in the interior of our sun) the probability is very high.

We have known the real energy source of the sun for almost fifty years, although the details came to light only within the last twenty years and some questions remain unanswered. But we do know at least that the sun obtains its energy from nuclear transformations. The process is that of nuclear fusion, the build-up of atomic nuclei of higher atomic numbers from elements of lower ones. In our sun it is the formation of helium from hydrogen.

Nuclear transformations are the processes during which millions of times more energy is liberated than during chemical reactions. Compared with nuclear transformations the energy liberation of the most powerful explosive is negligible. Such nuclear transformations are nevertheless comparable—except when triggered spontaneously in an atom bomb—rather to the smouldering of a pile of coal than to an explosion. Only their huge scale makes them such an enormous source of energy for the sun.

During nuclear transformation heavy atoms are split into lighter ones, or light atoms are fused into heavier ones. The first reaction is called nuclear fission; it occurs in the atom bomb and in nuclear power stations—in the bomb spontaneously and uncontrolled; in the power station in a slow, controlled process. Of course, this does not mean that a nuclear power station could be readily turned into an atom bomb either by design or by accident without a great deal of work. In nuclear fusion light atomic nuclei are fused into heavier ones, a process we have so far been able to imitate only in explosives.

But the life of the deuteron, or heavy hydrogen, nucleus is very short. On average, it captures a free proton, a hydrogen nucleus, after as little as six seconds, thereby becoming a helium nucleus according to the following equation.

$$H_1^2 + H_1^1 \rightarrow He_2^3 + \gamma$$

In this equation He = helium, and γ = short-wave electromagnetic gamma radiation liberated in the reaction.

If an He_2^3 nucleus collides with another He_2^3 nucleus, which is highly probable, the result will be a "proper," or stable, helium nucleus of nuclear number 4 and two hydrogen nuclei.

$$He_2^3 + He_2^3 \rightarrow He_2^4 + 2\,H_1^1$$

Another possibility consists in this reaction.

$$He_2^3 + He_2^4 \rightarrow Be_4^7 + \gamma$$

Here the fusion of helium isotope of nuclear number 3 with a stable helium nucleus of nuclear number 4, produces an atom of the element beryllium. The nucleus of this atom contains four protons and three neutrons. The isotope of beryllium is radioactive; it decays at a half-life of 43 days; that is, within this period half of a given quantity of it is transformed into lithium, emitting a positron.

$$Be^7_4 \rightarrow Li^7_3 + e^+ \nu$$

The loss of a positron by an atomic nucleus is equivalent to the loss of a positive charge, whereby a proton becomes a neutron; the chemical nature of the atom changes, but the nuclear number remains the same. The lithium atom is also unstable; on average it captures a proton after one minute, decaying as it does so into two helium atoms:

$$Li^7_3 + H^1_1 \rightarrow 2\,He^4_2$$

Thus, a new helium atom has been produced from four hydrogen atoms with the addition of a helium atom, which finally emerges intact from the chain of reactions.

The second sequence of reactions that qualifies as an energy source of our sun is the carbon cycle, also called the Bethe-Weizsaecker Cycle. It posits the presence of another element, carbon or nitrogen. The additional element acts as a catalyst, that is, as a substance whose presence is necessary to induce a reaction but which will emerge intact from the process—like the helium atom in the H + H reaction.

The Bethe-Weizsaecker cycle is a closed cycle; we can begin the observation of the reaction sequence at any stage. Let us begin with the reaction in which a carbon nucleus captures a hydrogen atom and thereby becomes a nitrogen isotope of nuclear number 13:

$$C^{12}_6 + H^1_1 \rightarrow N^{13}_7 + \gamma$$

Inside the sun, this reaction occurs for each C atom on average every thirteen million years; but the result, the nitrogen atom of nuclear number 13, is radioactive. After ten minutes this atom is converted into a carbon isotope of nuclear number 13, emitting a positron and a neutrino:

$$N^{13}_7 \rightarrow C^{13}_6 + e^+ + \nu$$

The resulting carbon nucleus 13 captures, on average after 2.7 million years, an additional proton (a hydrogen nucleus) and becomes an ordinary, nonradioactive, nitrogen 14:

$$C^{13}_6 + H^1_1 \rightarrow N^{14}_7 + \gamma$$

Here, too, a γ particle is emitted. Another proton capture converts nitrogen 14 into the oxygen isotope 15:

$$N^{14}_{7} + H^{1}_{1} \rightarrow O^{15}_{8} + \gamma$$

This proton capture occurs on average after 320 million years. Such an oxygen isotope, however, decays into a nitrogen atom, emitting a positron and a neutrino:

$$O^{15}_{8} \rightarrow N^{15}_{7} + e^{+} + \nu$$

The nitrogen now captures another hydrogen atom and decays into a carbon and a helium nucleus:

$$N^{15}_{7} + H^{1}_{1} \rightarrow C^{12}_{6} + He^{4}_{2} + \gamma$$

If we add up all the reaction times of this cycle we find that it takes 336 million years in the sun. This is, of course, an enormously long time; the sun manages its huge energy output only through a very large number of these reactions occurring simultaneously—new cycles start and others complete virtually all the time. To initiate this cycle at all an appropriate number of carbon and nitrogen atoms must have been present in the sun from the very beginning. One can calculate that they must have amounted to about 1 percent of the sun's weight, and this figure has been confirmed by spectrum analysis of the sun.

These reactions proceed in the way they do only on the basis of the physical magnitudes of pressure, temperature, and chemical composition that do in fact prevail inside our sun—which strongly confirms the theory of this process of energy production. Let us therefore examine the internal structure of the sun.

A LOOK AT THE INTERIOR OF THE SUN

It is impossible physically to look at or measure the interior of the sun. Yet, on the basis of the reactions and conditions we can observe on the surface, and of theoretical models we derive from them, we do know some facts about the internal structure of the sun.

The sun is an enormous sphere of gas with no clear-cut layers and boundaries; there are no abrupt transitions such as between air and water or air and solid ground on the earth. But certain "layers" of the sun are described: the solar corona (the outermost atmosphere of the sun), the chromosphere (the transitional stratum below the sun's "surface"), and the photosphere (the luminous stratum). These are violently churning masses of gas, so thin that on earth we would regard them as a vacuum. In spite of its huge extent, the photosphere is no more than 500km deep.

The photosphere is the feature of the sun that appears to be luminous to the naked eye and is regarded as the surface of the sun. The average density of the whole sun is 1.4g/cc (1.4 times the density of water), only about 25 percent of the earth's average density (5.5g/cc). The gas that forms the photosphere on the other hand, has a density of only 10^{-6} to 10^{-7}g/cc, or one millionth to one ten millionth of the density of water; this corresponds to less than one thousandth to one ten thousandth of the density of air at ground level on earth.

The further we move out into the chromosphere and the corona, the more the density of the gas decreases. In the chromosphere it diminishes with increasing altitude from 10^{-8} to 10^{-13}g/cc. In the solar corona it is further reduced, reaching density values that we cannot achieve even with our best vacuum pumps: 10^{-16} to 10^{-17}g/cc. At the far reaches the corona, (with density values of 10^{-17}g/cc) gradually changes into the interplanetary gas consisting partly of the solar wind which is continually emitted by the sun and spreads throughout the entire solar system. In the vicinity of the earth the density of solar wind is about 5 protons and 5 electrons per cubic centimeter; these particles travel at a velocity of about 500 km/sec.

The farther we descend from the surface of the sun toward its center, the more the gas pressure increases; so does the temperature. In the core of the sun (which is about the size of Jupiter), the density is 160g/cc (about twenty times the density of the Earth), and the temperature about 15,000,000°C. These are the conditions necessary for nuclear fusion processes to take place. The energy liberated during nuclear transformations takes the form of extremely short-wave, invisible gamma radiation. But at the high pressure and in the densely-packed matter of the center of the sun the gamma rays do not travel very far; after a short distance they are absorbed by other atoms and reemitted, a process that continuously recurs. Thus, the energy originating in the atomic nuclei in the form of gamma rays is transported to the surface of the sun along intricate paths. The more often a gamma ray is absorbed by an atom and reemitted, the more energy it loses, and its wavelength progressively increases. The original gamma ray gradually becomes an x-ray, an ultraviolet ray, and finally an electromagnetic wave of the length that we can perceive and therefore call visible light. There are so many absorptions and reemissions—reradiations of the individual "energy parcels" along this path—that it takes energy almost 50 million years to travel from the core to the photosphere. From here its progress is unhindered and it reaches the earth in only 8.33 minutes.

During this huge nuclear energy "combustion" the sun uses up about 5 million tons of hydrogen every second, and it has done so continuously for 4 or 5 billion years. Still, only a small fraction of the sun's hydrogen content has been converted into helium. At a steady rate of consumption the remaining hydrogen would last the sun another 100 billion years. It will, however, not come to this in practice; long before all the sun's hydrogen is burned up into helium the transformation process will have been disturbed by helium "slag" and have become unstable. But this disruption is in the remote future; we expect it to occur in about 5 to 7 billion years. The disruption process of the sun and other stars is discussed later in the book.

THE FACE OF THE SUN

Galileo, the first person to point a telescope at the sky, observed the sun, recognized the sunspots as a solar phenomenon, and began a systematic observation of these spots. Since then the sun has been under more or less continuous surveillance; events on its surface have been followed, noted, recorded, and, in more recent times, photographed and analyzed according to the state of the astronomical art. But only during the last few decades—thanks to new methods and possibilities of observation—have we been able to acquire an understanding of solar reactions and their interrelations with other bodies in the universe.

This was because previously we had to look at the sun in an extremely "one-sided" way; we were strictly limited to the possibilities of optical observation, the investigation of the sun in the visible region of the spectrum, and even this was severely restricted. We

The solar telescope erected on Kitt Peak, Arizona, in 1962, at an altitude of 2,064m (where "seeing" is particularly clear). The diameter of the primary mirror of this McMath solar telescope is 150cm; its focal length is 91.4m. To date it is the world's largest solar telescope. A mirror on the vertical column directs the sun's rays through a sloping structure into basement rooms, where the sun is photographed and investigated in various regions of the spectrum.

Facing page:
Top: The large dome of the solar observatory on the Wendelstein mountain in Bavaria. The smaller, lower telescope is used for observation of the sun in integral light (ordinary daylight); the longer, upper one is a coronagraph, with which the corona and the protuberances are investigated.
Bottom: The solar telescope of the High Altitude Observatory (a joint facility of Harvard University and the University of Colorado in Boulder).

had to wait for modern observational instruments, for the expansion of the observed solar spectrum by means of radio-, ultraviolet-, and x-ray astronomy and for the techniques of observation from space stations, artificial satellites, and space probes. All the space probes, satellites, and space stations notwithstanding, observing and tracking the sun from ground level has lost none of its previous interest. On the contrary, the various new disciplines have no more replaced the former earth-bound solar research than space astronomy replaced earth-bound astronomical research—in fact, these various branches of astronomy supplement each other.

Today an estimated 20 percent of all astronomers, a total of about 500, specialize in solar research. Many solar observatories have highly sophisticated equipment; special solar observation facilities exist on Mount Wilson, California; on Kitt Peak, Arizona; on Wendelstein, Bavaria; on the Crimea, in the Soviet Union, at the Tokyo Observatory; and on Pic du Midi, in the French Pyrenees, to mention only a few. The community of solar researchers extends around the globe and solar surveillance is carried on practically nonstop. Added to this is radio observation and observation of terrestrial phenomena dependent on solar reactions, such as the fluctuations of the earth's magnetic field, interference with short-wave radio transmissions, the aurora borealis, and other events in which relations with solar activities are suspected. This activity of the sun is subject to changes. It can be measured by the number of visible sunspots and by phenomena in the corona and in the ultraviolet and radio emission of the sun (for instance, in the 10.7cm wave band); and it undergoes long-term periodic fluctuations. The most outstanding period is eleven years; every eleven years many sunspots are visible, and few are visible in the years in between. Needless to say this does not mean that we can see sunspots only during the times of sunspot maxima, or that the sun is completely free from spots during a sunspot minimum. A com-

---

The solar probe Helios being checked in the workshops of Messerschmitt-Boelkow-Blohm (the main contractor) near Munich. On the right is a half-extended instrument boom. Two probes of this type were sent into orbit around the sun in 1974 and 1976. In their elliptical orbits they approached the sun as closely as 46.5 and 43 million kilometers, respectively; and they transmitted a multitude of data about space near the sun.

parison of the mean values of weeks or months with each other results in a curve that clearly reflects the eleven-year sunspot cycle.

The Sunspots

Several methods are available with which we can demonstrate the extent of solar activity. One of the oldest and most reliable is the Zurich Sunspot Number introduced in 1849 by Zurich astronomer Rudolf Wolf and still used today in spite of several ingenious rival methods.

To arrive at the Zurich Sunspot Number the number of *groups* of sunspots visible on the sun is determined, multiplied by 10, and the number of *individual* spots visible on the sun is added to the result. This produces the relative number, a measure that enables us to provide continuous information about the activity of the sun through a telescope.

But the relative number obviously differs from telescope to telescope and from observer to observer. In addition to the size and therefore performance of a telescope and the local seeing conditions, the visual acuity and the standard of evaluation of the individual observer also play an important part. To establish a worldwide standard of comparison the telescope with which Wolf began his solar observation was declared the international standard telescope more than a hundred years ago. All solar observations since have been referred to this refractor of 8cm aperature and 110cm focal length. The individually obtained Zurich Numbers are therefore multiplied by a factor determined and passed on to the individual observers by the Zurich observatory. The result is the relative number one would obtain if the reported observations had been made through Wolf's telescope in conditions of clear weather in Zurich. The Zurich Sunspot Number R follows the equation $R = k\,(10g + f)$. In this equation k = the individual instrument factor; g = the number of sunspot groups; and f = the number of the visible individual sunspots.

For a long time the nature of sunspots was a mystery. Observation has shown that they suddenly appear as small spots on the sun's disk. A spot may remain unchanged; it may disappear completely after a few days; other spots emerge in its vicinity. Sunspots may become larger and develop into a sunspot group, which may persist for days, weeks, and occasionally even months, moving from one side of the sun's disk to the opposite as the result of the revolution of the sun's gaseous sphere on its axis. Because the period of revolution of the sun (observed from the earth) is more than twenty-seven days, a sunspot that emerges on the left-hand edge of the sun's disk moves to the right-hand edge and disappears within a period of fourteen days. Large, persistent groups can be seen to reemerge at the eastern edge thirteen to fourteen days later, after their move across the far side of the sun. Some groups survive several revolutions of the sun and reemerge repeatedly before dissolving.

The dimensions of the sunspots are considerable. The smaller ones are about as large as the continents on earth; the medium-sized to large ones can be as large as the earth or Jupiter. Their structure

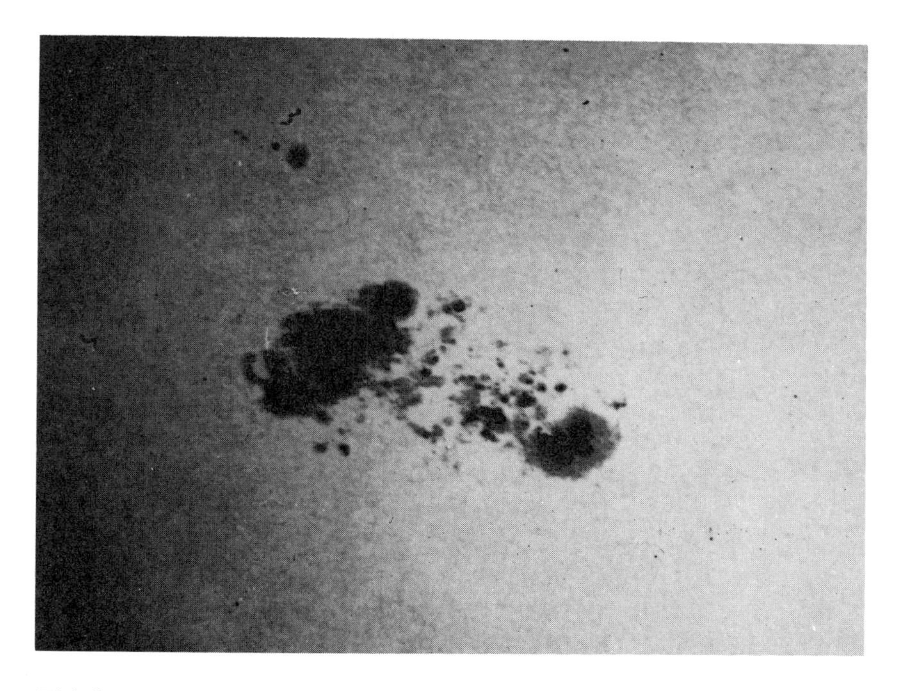

This large group of sunspots was photographed at the Wendelstein solar observatory on April 6, 1947. The picture shows very clearly the rich details of sunspots of various sizes. Penumbra and umbra are readily recognizable.

is extraordinarily complex. The most prominent distinctions are a dark, inner part—the umbra (shadow)—and a semidark surrounding area—the penumbra (half-shadow). Sunspots are the result of enormous magnetic fields set up by reactions that probably occur below the photosphere but are not yet understood. These fields are extremely strong and in extensive groups reach about 5,000 Gauss. (For the sake of comparison, the normal "undisturbed" magnetic field of the sun has a strength of 1 to 2 Gauss; that of the earth, 0.31 to 0.58 Gauss.) These enormous magnetic fields of the sun provide the explanation of almost all the phenomena we can observe in the sunspots. The darker hue is due to their lower temperature, which in turn is the result of the inhibiting effect of strong magnetic fields

The curve of Zurich Sunspot Numbers from 1700 to the present.

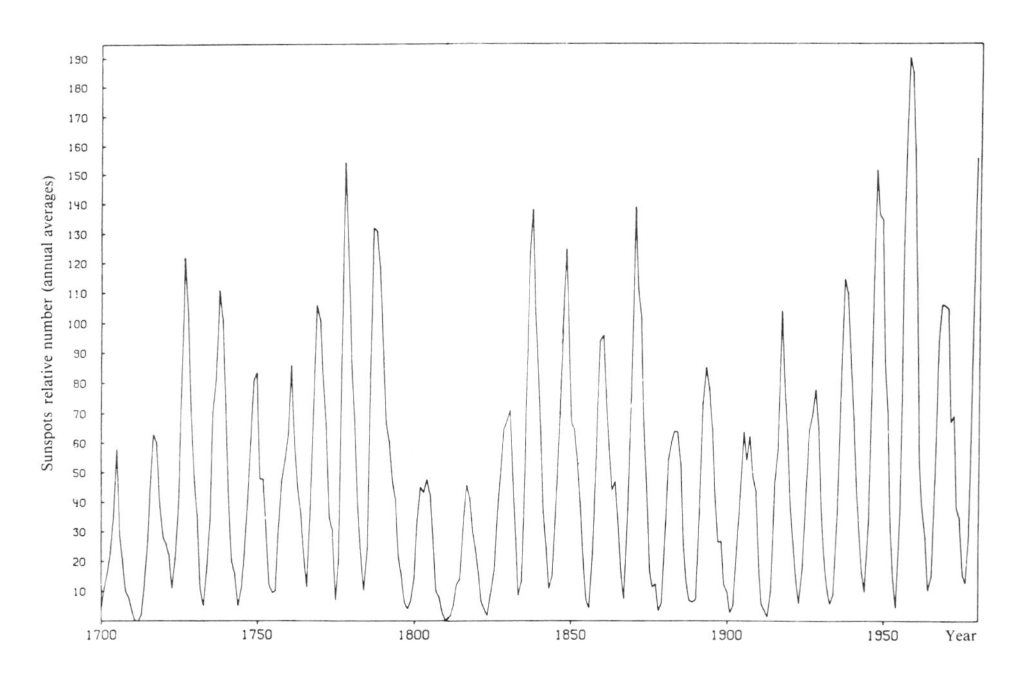

This shows not only the average eleven-year fluctuations of sunspot activity but also the very high and very low sunspot maxima. Many connections between the sunspot curve and terrestrial events have been proved; others are suspected.

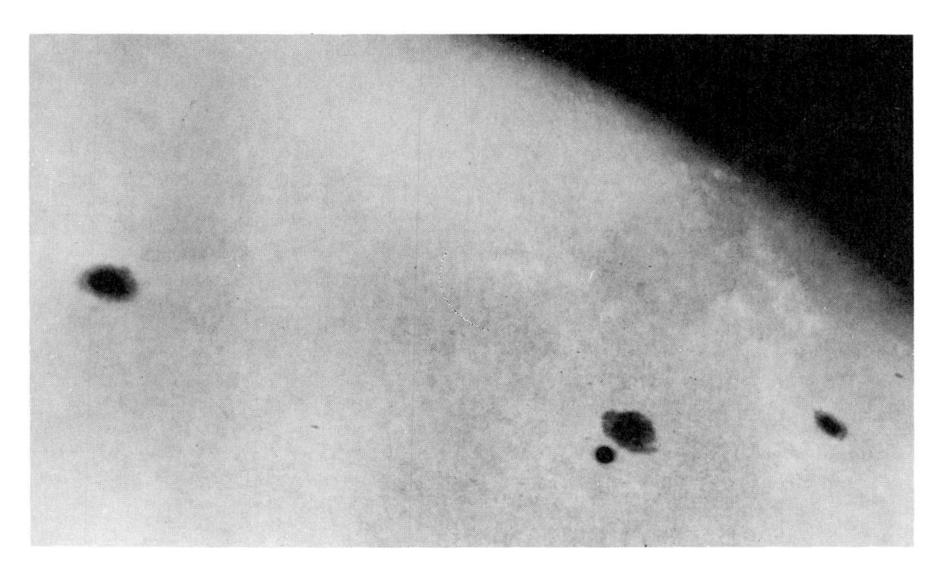

This picture of the sun from November 1970 shows not only sunspots but also the tiny round disk of Mercury silhouetted against the sun. Such transits of Mercury and Venus were often used to measure the distance of the sun by taking precise bearings of the planet against the sun's disk.

on the normal heat flow to the surface of the sun. Arches and bridges observable in the sunspots precisely follow the pattern of the magnetic field lines, and the differential magnetic polarization of the individual sunspots is also explained by the magnetic fields.

The darkness of the sunspots is only relative. Placed in the sky on its own, such a sunspot would still be 1,000 times brighter than the full moon (the sun itself is 1.2 million times brighter). Again, the sunspots are darker than the undisturbed surface of the sun because their temperature is lower. Whereas the surface temperature is not quite 6,000°C, that of the sunspots is no more than 4,000 to 4,500°C. The sunspots are "refrigerating machines" in which huge movements of gases take place.

The sun also contains hotter spots, most of which are found in the neighborhood of the sunspots—the faculae. Faculae have a temperature of about 7,000°C, and they rise from the photosphere as superheated masses of gas. When occurring without sunspots in the vicinity they often precede the development of sunspot groups, indicating that the surface of the sun is disturbed in this area.

Sunspots were, on occasion, observed long before the invention of the telescope, although the sightings were limited to the largest groups of spots. We have reports from China and the East going back about 2,000 years of an average five to ten sightings of sunspots every century. Large sunspots are indeed easily seen when the intense light of the sun is filtered by fog or smoke, or when (shortly after sunrise or before sunset) sunlight has to pass through layers of haze close to the horizon. Mostly, however, the sunspots are mistaken for birds, unknown planets, or other independent objects passing across the sun. Galileo was the first to suggest that sunspots were objects on the sun's surface. A violent quarrel erupted between Galileo and David Fabricius, Thomas Harriott, and Christoph Scheiner over the priority of the discovery of the sunspots.

The sunspot curve—representing the frequency of sunspots as a function of time—has been successfully traced, with a few initial gaps, back to 1610, and continuously to about 1700. Therefore, we have information about the extent and occurrence of more than thirty sunspot maxima and the intermediate minima. This reveals two facts: there are maxima on average every 11.2 years, but this period varies between 17.1 and 7.3 years. The height of the individual maxima also varies. Outstanding maxima were observed in 1778, 1788, 1837, 1870, and 1957. The last-mentioned sunspot maximum was the highest ever recorded. But there were also periods of few sunspots; in fact there is a suspicion that no solar activity took place at all during the second half of the seventeenth century. By and large we can, however, say that the roughly eleven-year sunspot cycle has been quite constant throughout the centuries.

But it is by no means settled whether it would not be more accurate to speak of a twenty-two-year rather than an eleven-year cycle. It has been shown that the magnetic polarity of the sunspots of the two hemispheres is reversed every twenty-two years. If during a sunspot period the preceding spot on the Northern hemisphere is a magnetic North Pole and that on the Southern hemisphere a South Pole, the conditions reverse with the subsequent sunspot period; the preceding spot on the Northern hemisphere displays south polarity, and vice versa. With the change to the succeeding period the polarities of the sunspots also change so that—at least from the aspect of magnetic activity—we seem justified to speak of a twenty-two year period.

Other long-term periods obviously superimposed on the eleven-year sunspot cycle have been observed. A period of eighty years is particularly noteworthy. Abnormally high maxima appear to occur every eighty years—the most recent in 1957.

The cause of all the periodicities of the sunspots (like all the other forms of solar activity) has not yet been established; the reactions evidently proceed below the photosphere and are connected through a cyclic mechanism of the differential revolution of the sun. The rate at which the gaseous sphere of the sun revolves on its axis is not uniform; the areas at high latitudes revolve more slowly than the equatorial belt. The difference is considerable. The sidereal period of revolution of the sun's equator is 24 days 6 hours; that of the sun's poles is about 35 days. But how the mechanism of the periodicities of the sun functions in detail is still a subject of research by theoretical astrophysicists.

Careful observation of the sun's surface through a telescope shows a granular structure in continuous motion. These cellular

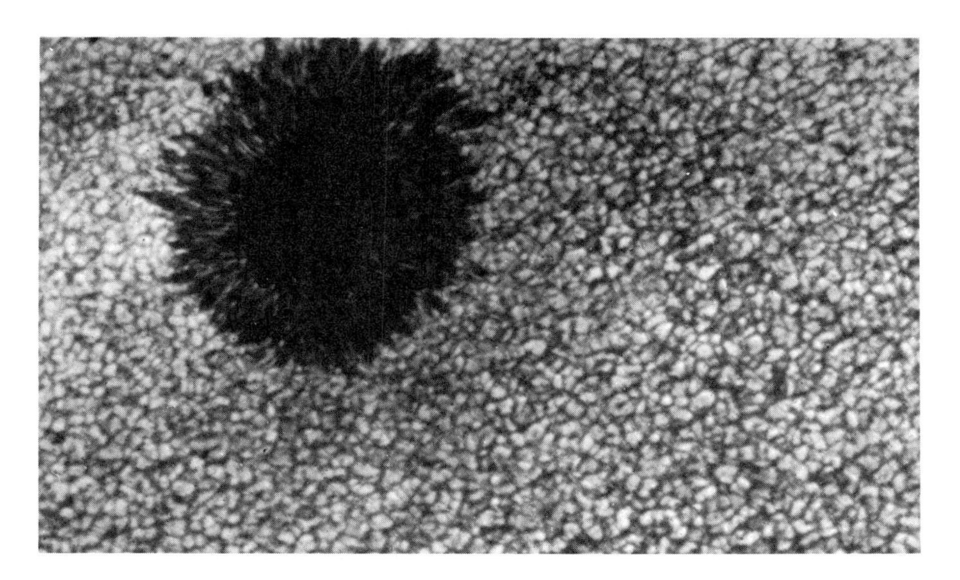

Note how disturbances in the earth's atmosphere impair the quality of photographs of the sun. This was taken on July 30, 1970, by a Soviet research balloon 20km above ground level and clearly shows the granulation of the sun's surface.

features, which make up the photosphere, are called granules and resemble the polygonal cells of a honeycomb. Ranging in diameter from 500 to 1000km and often compared to boiling rice grains, the granules have an average life of only five minutes. They are about 300°C hotter than their darker surroundings; they cover about one third of the entire photosphere; and they are caused by convection. Like turbulent towering clouds or bubbles in a boiling liquid they move to and fro, convey hot gases from deeper layers of the sun upward, radiate the conveyed heat into space, and subside. They move vertically at a rate of 18km per minute. In favorable conditions this granulation of the sun can be seen even through small telescopes. Better insight into granulation has, however, been acquired only within the last ten years through cameras on high-altitude research balloons.

THE UNKNOWN SUN

Many reactions on the sun can be observed from earth only with the aid of ingenious techniques; others cannot be observed at all. Among those that can be observed are the chromosphere, the corona, the solar radio emission, the protuberances, and some solar energy eruptions; among those that cannot are the appearance of the sun in x-ray "light" and in other very short-wave radiations of the spectrum that are unable to penetrate the atmosphere of the earth.

The Chromosphere

The chromosphere is a layer 2,000 to 4,000km thick above the sun's photosphere; in the past it could be observed only during the rare total eclipses of the sun, when the new moon covers the glaring photosphere, making the less bright layers above it briefly visible. It is called the chromosphere because it appears as a reddish fringe shortly before and shortly after total eclipse. The layer consists of ionized hydrogen. It conducts electricity outward, and its atoms have emitted an electron each. Distinctive features of the chromosphere are its continuous motion and the absence of a smooth upper boundary. Spicules—numerous bulges and masses of gas shooting pointedly upward—form the transition zone to the corona. These spicules sometimes penetrate the corona to a depth of 5,000 to 15,000km. They rise and collapse within five to ten minutes, comparable to spindrift in a stormy sea.

The corona is a plasma—a gas consisting of atomic nuclei and free electrons. The ionization is the result of the high temperatures in this region. Starting from the lower chromosphere, which has a temperature of about 7,500°C, the temperature of the corona rises to 1 to 2 million degrees C. As previously mentioned, however, at 10^{-16}g/cc the density of the gas is extremely low. Turbulence and the mechanical motion of atoms cause the high temperatures in this region.

The higher temperatures in the solar atmosphere than in the photosphere must at first appear puzzling. It is reasonable to assume that the temperature would drop with increasing distance from the photosphere, in the same way as it does the farther away we move

This picture of a total solar eclipse was taken on May 30, 1965, from a NASA research aircraft 12km above the southern Pacific. The structure of the corona is clearly revealed at this altitude.

from a stove. This apparent paradox is explained by the photosphere being no *ordinary* emitter of energy and the conditions of the corona not being determined solely by radiation energy. Kinetic energy or reactions in the granules, the spicules, and other undulatory phenomena of the solar gas also contribute. Part of this energy of the photosphere is thus mechanically transferred into the chromosphere and into the corona, where it manifests itself as thermal energy, in the motion of atoms and atomic building bricks. Another part of the energy of the photosphere, radiation energy, traverses the turbulent layers of chromosphere and corona unaffected by the reactions around it and proceeds into space.

As already mentioned, the chromosphere and the corona could in the past be observed only during the brief minutes of a total eclipse. This situation has now changed thanks to the spectroscope and the spectrograph. A spectroscope is basically a triangular glass rod, a prism that disperses sunlight into a band of the colors of the rainbow. Sir Isaac Newton explained the function of the prism. He demonstrated that white light is the sum of all the colors of the rainbow and the spectroscope disperes it into its "components."

Following in Newton's footsteps; William Wollaston conducted experiments in 1802 to improve this method. He placed a narrow slit in front of the prism of his spectroscope, through which the sunlight had first to pass. He thereby achieved an improved, higher resolution and discovered that a spectrum obtained in this way displays a number of dark lines vertically crossing the band of colors. Joseph von Fraunhofer, a Munich optician, repeated Wollaston's experiments and found more than 500 such dark lines in the spectrum of the sun. Today they are called Fraunhofer Lines. Gustav Robert Kirchhoff and Robert Wilhelm Bunsen realized that the Fraunhofer Lines indicate certain chemical elements. Spectrum analysis thus became a method with which we can determine the gases that glow in the sun and in other stars. From the intensity of the Fraunhofer Lines we can deduce the quantities of certain elements present in the sun or in a star. A spectrum also provides information about movements and changes in movements, which show up as displacements and broadenings of the spectrum lines. The spectroscope thus proved to be the astronomers' and physicists' magic wand. Information suddenly became available about the sun and the stars that exceeded scientists' wildest dreams.

The spectroscope also enables us to investigate strata at various altitudes above the sun. For instance, if it is directed at the photosphere it shows the dark Fraunhofer Lines—absorption lines—on a continuous spectrum of colors, a continuum. Such a continuum, for instance the rainbow, indicates whether the light-emitting body is a solid or a liquid.

The radiation of the sun and of the stars produces basically a continuum simply because the density of their interior is very great. But the light has to pass through the thinner and cooler outer layers of the stars which absorb the wavelength ranges they would themselves emit as luminous gases if they were hotter. The result is an absorption spectrum, in which the lines of those elements appear which are present in the cooler gas on top. If these gases existed without the solid or liquid body, they would produce, narrow, vertical, bright lines in exactly the same sites in an emission spectrum. This strange, complex behavior of the spectra is explained by the atomic structure of elements; it is connected with the fact that the electrons, which orbit the atomic nuclei, describe different paths of different energy levels. If the electrons jump from an inner to an

Many Fraunhofer Lines can be seen in this picture of the solar spectrum. They have all been accurately cataloged and are identifiable. This is the spectrum of a flare, a sudden energy eruption on the sun. The picture was taken on September 18, 1957, at the Sacramento Peak Observatory. A "flare patrol" for the continuous monitoring of the sun has been in existence for some years.

outer orbit at a higher energy level, they absorb energy. If they jump from an outer to an inner orbit, they liberate energy. But these jumps occur only in discrete steps or quanta. If an energy jump takes place from outside toward the center, the same amount of energy is liberated as will be absorbed during a jump in the opposite direction. This explains the inversion of emission spectrum lines to absorption spectrum lines, or Fraunhofer Lines.

So far 20,000 lines have been discovered in the spectrum of the sun. These lines are often, but wrongly, described as being "thin as a hair." Their width is measurable, and their intensity varies across it. These line widths and intensities provide, although in a very complex way, information about the proportion of a chemical element relative to others. Pressure and temperature conditions, as well as magnetic and electric fields, also affect the width and appearance of these lines.

With the aid of a spectroheliograph, pictures of the sun can be obtained within the region of a single spectrum line. The spectroheliograph is a piece of apparatus in which a slit selects a narrow strip from the image of the sun produced by a telescope. The light passing through the slit is dispersed into a spectrum by means of

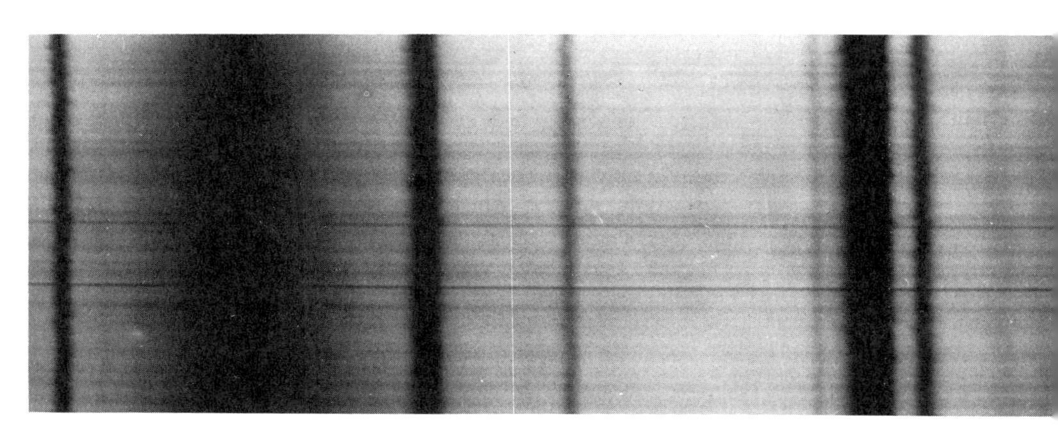

This is part of a spectrogram taken at the Sacramento Peak Observatory. High dispersion must be achieved for the recording of fine details of the spectrum lines. Today more than 20,000 lines of the solar spectrum have been identified and cataloged. The nine most prominent are lettered A through K, beginning at the red end of the spectrum.

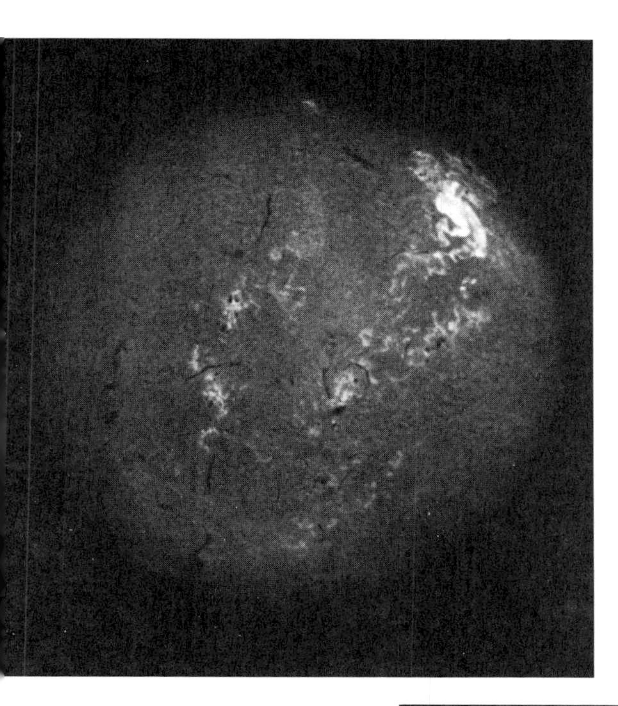

The sun in the light of hydrogen, photographed at the Sacramento Peak Observatory on May 10, 1959. Note the flare in the top right.

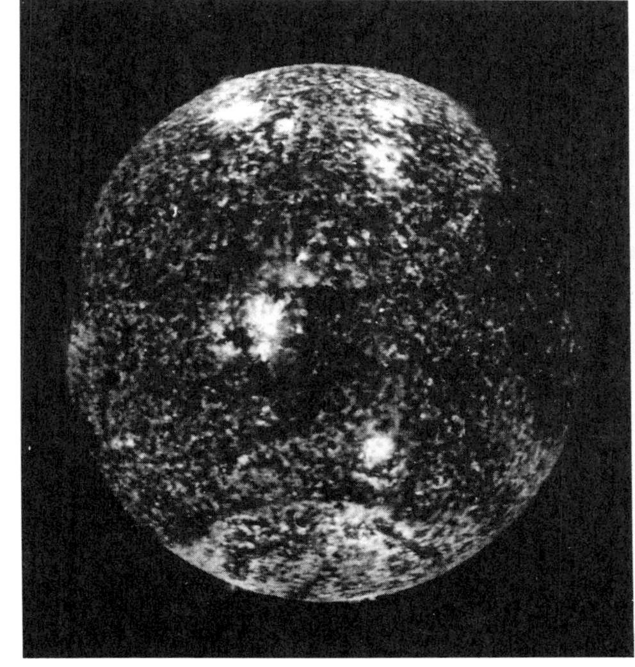

The sun in the light of ionized helium, photographed by Skylab on August 2, 1973, showing the chromosphere in the temperature range of 50,000°C.

prisms or diffraction gratings. A second slit then selects a single line from the whole spectrum. The instrument successively scans the entire sun with the first slit, thus producing an image of the sun composed of many strips in monochromatic light, light of a certain element. Depending whether the slit of the spectroscope is pointed at the center or at the marginal portions of the lines, photographs can be taken of layers at different altitudes above the sun. Visual observation is possible if, instead of a spectroheliograph, a spectrohelioscope is used. Thus, we can move between the photosphere and the chromosphere simply by making various instrument adjustments.

The inventor of the spectrohelioscope was the twenty-one-year-old George Ellery Hale. During his studies in 1889 he changed a spectrograph at the Harvard Observatory so that it formed an image not only of a narrow strip of the sun in a single line of the spectrum, but also reproduced contiguous strips of the sun's surface on a moving photographic plate.

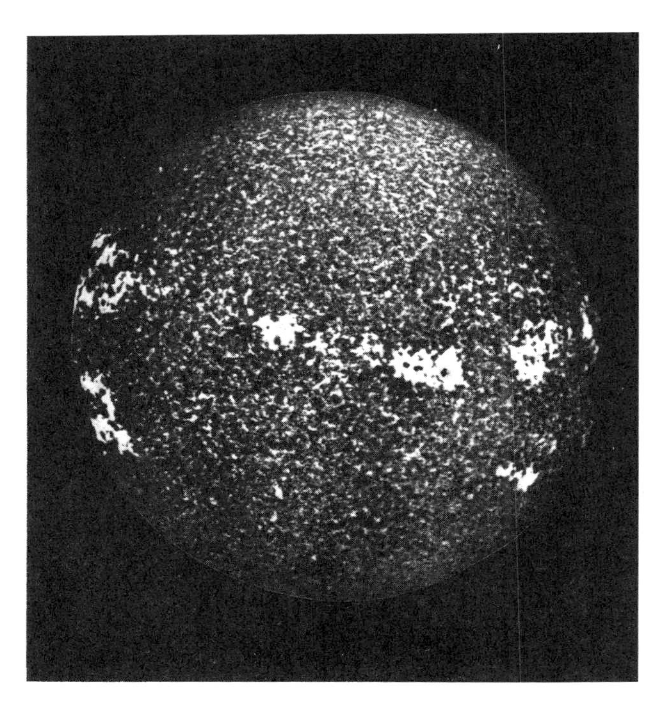

Spectroheliogram of the sun in the light of the H-line of calcium. With the technique of photographing the sun in the light of selected regions of the spectrum, various levels and temperature strata can be reached.

During the 1930s filters that transmit only very narrow regions of the spectrum were developed, thereby producing the same effect as the spectroheliograph and the spectrohelioscope without the need for complicated mechanical equipment. Such solar filters of selected regions of the spectrum—for instance the famous hydrogen line H_α—are now available even to amateurs, so that every owner of a small telescope and such a filter is able to examine monochromatic images of the sun in clear weather.

With the aid of the spectroheliograph, spectrohelioscope, and filters many previously unknown reactions on the sun have been "brought to light." One early discovery made on spectroheliograms was of faculae, luminous bright areas above the groups of sunspots or above areas where such groups can be shortly expected in the chromosphere. They consist of hot, relatively dense gases, which are better indicators of solar activity than sunspots. In addition, they form the centers of the sites where chromospheric eruptions flare up from time to time. These sudden outbreaks of radiation last from a few minutes to one to two hours, and most can be observed only with the spectrohelioscope. They are the most intensive solar reactions we know. They can be seen above all on monochromatic pictures, and mainly on those of the H_α line. Against the background of the integrated white light coming from the photosphere they can hardly be seen at all. Only one such chromospheric eruption, also called flare, was observed in white light before the invention of the spectroscope; since the introduction of the spectroheliograph and monochromatic filters a few have been noted against white light.

Flares can, however, often be observed with the spectroheliograph or through H_α filters. A "Flare Patrol" has been in existence at most important solar observatories for several decades for the continuous surveillance of the sun for chromospheric eruptions. This surveillance is partly semiautomatic, with photographs continuously being taken of the sun in the light of hydrogen (H_α), to cover every one of these energy eruptions from beginning to end (provided the sky above the observatory is clear).

This monitoring system is operated because solar energy eruptions have many effects on interplanetary space and on earth. Such an energy eruption on the sun is a strictly local event; the focus from which the energy is emitted is rarely larger than a few tenths of a percent of the surface of the sun. But the density is so enormous that during a major eruption of this kind the same amount of energy is liberated within a few minutes as would be by the explosion of 10 million hydrogen bombs. Only a small proportion of this energy is emitted in the visible region of light; the major part is radiated in the form of ultraviolet rays, x-rays, and particle rays. In addition, powerful emissions, the bursts or eruptions of radio emissions caused by plasma oscillations in the corona, occur in the radio wave range.

A small part of this energy reaches the earth, but it is still large enough to cause disturbances in the ionosphere, leading, for instance, to severe breakdown of short-wave radio communication. The eruption of x-ray and ultraviolet radiation cannot be detected from the surface of the earth, because these types of radiation are screened by the earth's atmosphere. During the first years of space

Skylab, at an altitude of 435km, in orbit around the earth. In the front section of the space station are the wing-shaped solar cell surfaces for energy supply and the Apollo telescope assembly for solar research. Evaluation of the 812,000 photographs of the sun taken from Skylab and of the data recorded is still proceeding.

research (since about 1957) it was possible to record the later stages of some x-ray eruptions with measuring instruments on high-altitude rockets. Later, solar research satellites, such as the Orbiting Solar Observatories (OSO), transmitted the first spectrographs of x-ray emissions of the eruptions to the surface of the earth. But we only received detailed information and pictures of entire x-ray spectra through the activities on the Skylab space station between May 1973 and February 1974. One of the tasks of the three crews (who spent twenty-eight, fifty-nine, and eighty-four days, respectively, on board the station) was to monitor the sun continuously for such energy eruptions. The astronauts took more than 182,000 pictures of the sun, about 40,000 in the x-ray region.

The solar research programs on Skylab have given us much deeper insight than before into events on the sun and into the effects of these phenomena on the earth, on its atmosphere, and on interplanetary space.

Chromospheric flares as well as energy eruptions occur in regions of the sun that are extremely active magnetically. The chromospheric flares are thus excellent replicas of the strong magnetic field that always surrounds them, and the chromospheric eruptions are connected with the magnetic fields of the sunspots below without, however, affecting the latter. It is now assumed that the eruptions are the result of a sudden conversion of magnetic energy into thermal, light, and kinetic energy. The outcome of the research carried out on Skylab also points to this explanation. We now know that the eruptions occurring in visible light are but a part of the life of a flare, and that its true significance and main activity precedes the appearance of the optical phenomena. We also know that the chromospheric eruptions, like so many other events on the sun, strictly follow the eleven-year sunspot cycle.

The Corona

Beyond the chromosphere lies the corona, the outermost source of its radiation, where the protuberances occur, mighty flames of hot gases shooting into space. Before the invention of appropriate research instruments, neither the chromosphere, the corona, nor the protuberances could be observed except during the short, rare, total eclipse of the sun. Compared with the normal sunlight of the photosphere they are so faint that the luminous disk of the sun renders them invisible to the eye.

For decades the corona was a complete mystery. On August 7, 1869, during a total eclipse of the sun, the spectrum of the corona was successfully observed for the first time. It was totally different from the well-known continuum with the Fraunhofer Lines that the photosphere presents. The spectrum of the corona consisted of a faint continuum, on which a few bright spectrum lines were superimposed. This by itself was not very exciting because it was understood that the appearance of the spectrum would change with the region of the sun observed.

Charles Young, the first astronomer to investigate the spectrum of the corona, discovered that the chromosphere has an entirely different spectrum—an inversion, as it were—of the normal photospheric spectrum of the sun. Where in the normal spectrum Fraunhofer Lines can be seen in a continuum, bright emission lines now appear against the background of a faint continuum. But the striking feature of the coronal spectrum is that the bright emission lines do not coincide with the Fraunhofer Lines of the ordinary solar spectrum. In the coronal spectrum Young had found, for instance, a strong green line that had no Fraunhofer "counterpart," in the photospheric spectrum, nor was there any known line in the spectra produced on earth that occupied the place of this green line. At first it was suspected that the green line indicated an as yet unknown element, which was duly named "coronium." This view was held until 1940, when it was found that the emission lines of the corona were those of known metals at extremely high temperatures, metals that were repeatedly ionized and had lost several of their electrons. This was the first indication of temperatures of 1 million degrees C and higher in the corona. The green corona line proved to be that of repeatedly ionized iron. Lines of even more highly ionized calcium, nickel, and other heavy elements were also identified.

Like almost all solar phenomena, the corona is subject to continual changes, which also reflect the eleven-year cycle of the sun. The corona can now be observed with the coronagraph, even without a total solar eclipse, but only from high mountains because of its extreme faintness. Its outer parts especially are not visible from the earth except during a total eclipse.

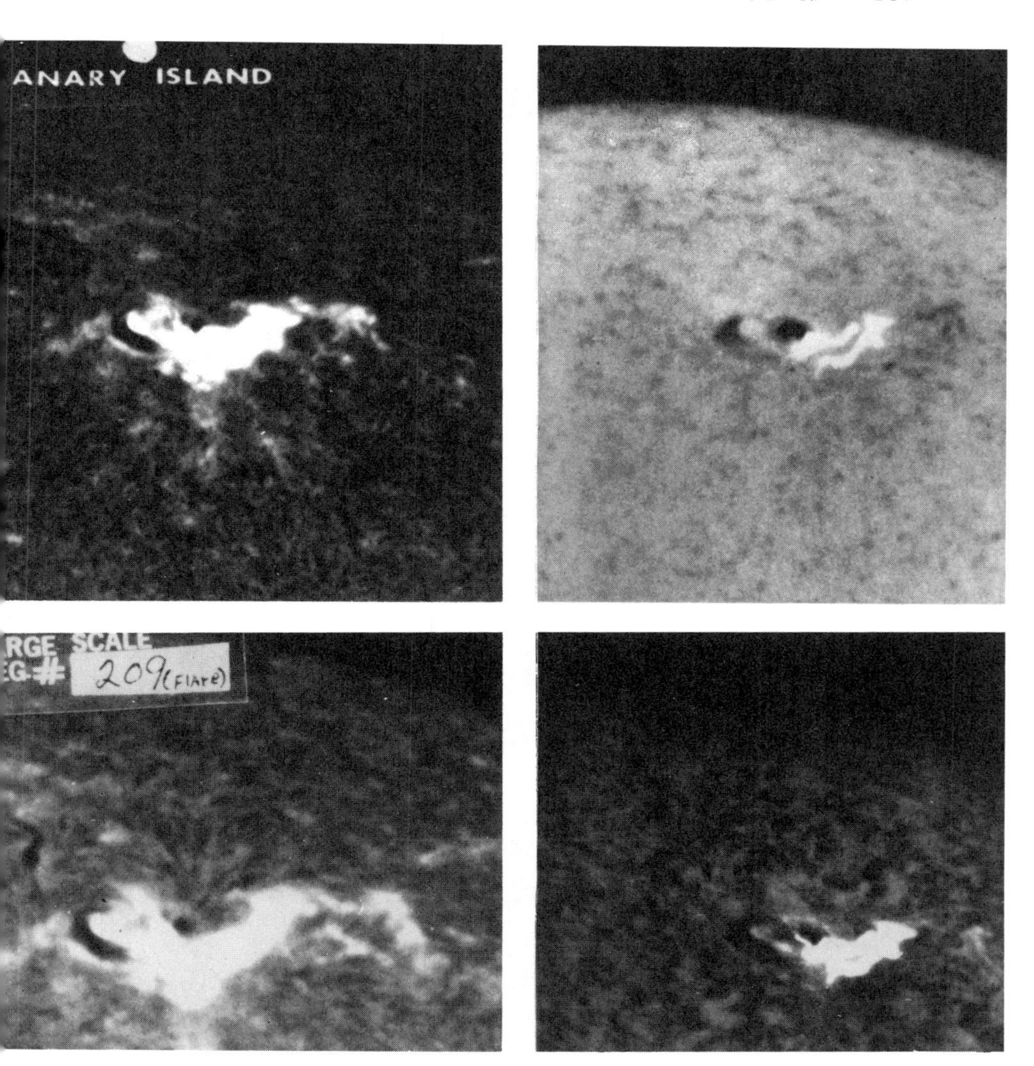

This, one of the largest solar eruptions, occurred on September 7, 1973, and lasted about two hours. A few minutes after it began, worldwide ultrashort-wave radio communication broke down. Such disruptions of communication are the effect on the earth of solar activity. These pictures were taken by Skylab. The bright area in the pictures equals seventeen earth surfaces. The two pictures on the right show the eruption in the light of the H-line.

The coronagraph, invented by Bernard Lyot in 1930, is an extremely precise instrument, consisting of an optical system as free as possible from scatter and containing a conical stop—a kind of artificial moon that obscures the photosphere of the sun. Lyot set up the first coronagraph on Pic due Midi in the French Pyrenees, where the disturbing stray light is extremely weak. Another coronagraph has been installed on Wendelstein in Bavaria. Skylab, needless to say, also had a coronagraph on board.

Not until the nineteenth century did the corona and the protuberances of the sun attract special interest. But research of these phenomena made little headway at first. Although two total eclipses of the sun occur every three years, they are extremely rare events in any given location. Also, the zone of totality is only a narrow strip never more than 300km wide, and often even narrower. In this area the moon's shadow races across the earth at an average speed of 580m/sec, almost 2,100 kmph or about twice the speed of sound. At any given point on the earth's surface a total eclipse never lasts longer than 7.6 minutes. And only during the phase of totality can the corona be observed. If any portion of the sun remains exposed its blinding light swamps the corona and the protuberances and we are unable to see them.

The regions of the earth affected by a solar eclipse up to the middle of this century could often be reached only after long and difficult journeys; and numerous reports, especially of early solar eclipse expeditions, read like adventure stories. Only occasionally will such a total eclipse be observable from, as we used to say in the past, a "civilized" region of the earth. In Europe in the twentieth

---

Facing page:
Top: A protuberance at the peak of its development. Its height—about 25,000km—is twice the earth's diameter. Some protuberances are even higher than this and escape in part from the sun into interplanetary space. This picture was taken at the Wendelstein solar observatory on June 11, 1948.

Bottom: A protuberance in the form of loops, photographed in hydrogen light. Such phenomena are relatively transient. This protuberance, recorded at the Mt. Climax solar observatory in the United States, persisted for less than two hours. The motions and shapes are triggered by intense magnetic fields limited to local areas of the sun.

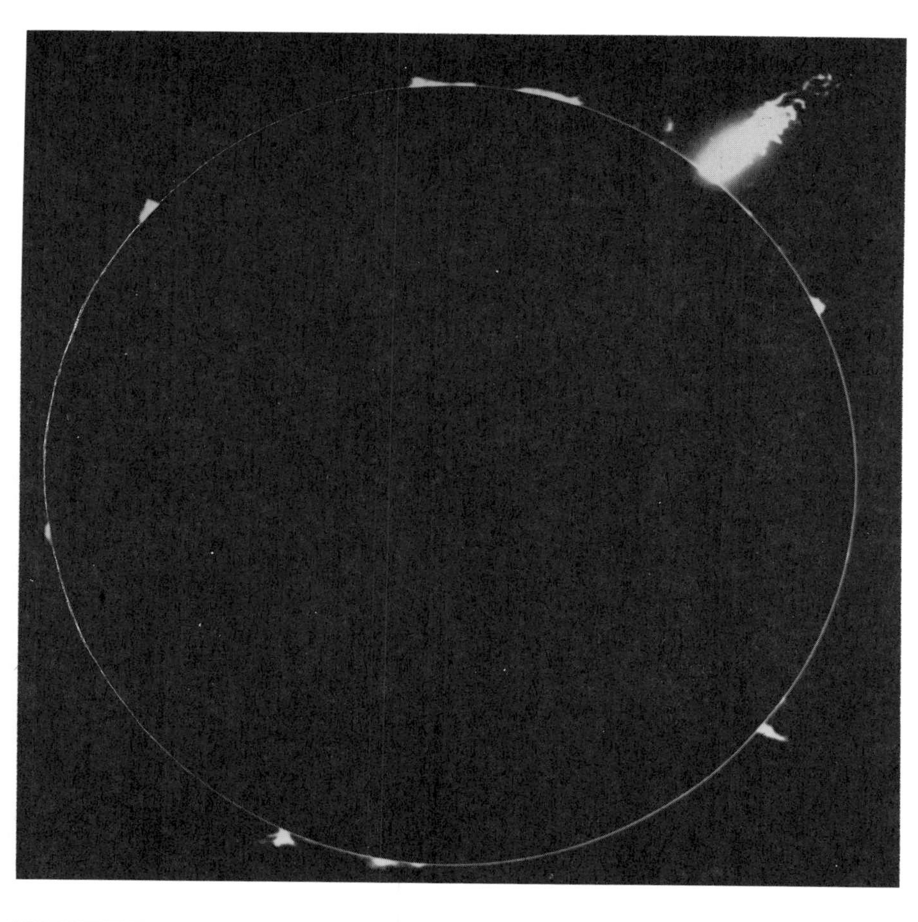

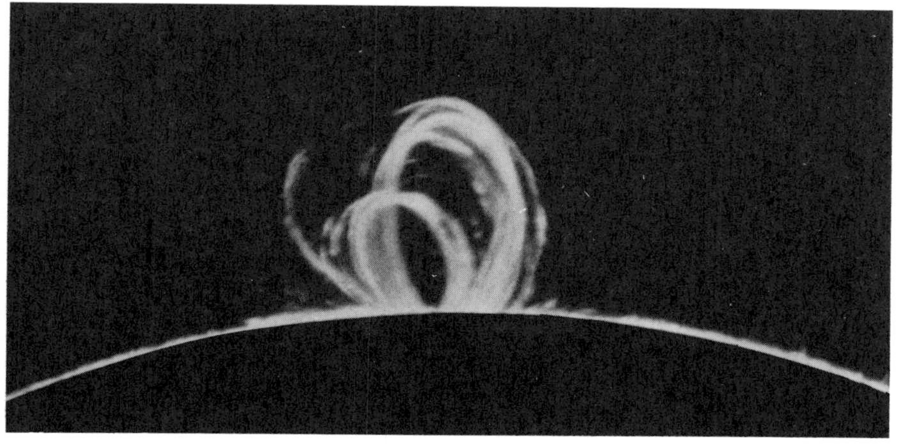

century, for instance, there were only ten total, two annular-total, and one annular eclipse of the sun. (Annular eclipses occur when the moon is at the apogee, its maximum distance from the earth. Its apparent diameter in the sky is therefore smaller than that of the sun and an annular zone of the sun will not be obscured.) Scientifically, annular eclipses are not as interesting as total ones. The last total eclipse in Europe occurred in 1961; the next will be in 1990; after that, in 1999, when, on August 11, its umbra will sweep through southern Germany and the western extremity of Cornwall, England. Only after the invention of the protuberance spectroscope, the spectroheliograph, and the coronagraph did the conditions of the exploration of these phenomena improve.

Even photographing the corona during the brief period of a total eclipse was difficult. Therefore, only drawings were made, but they depended so much on the individual skill and character of the artists—as a rule astronomers rather than draftsmen—that they proved rather worthless. The first photograph of the corona of the eclipsed sun was taken in 1851; the second was not obtained for another forty years. The reason was not only the inherent difficulty; the primitive quality of early photographic plates was equally responsible. Matters improved only toward the end of the last century when, in 1889, the Lick Observatory initiated a systematic program of coronal photography during total eclipse, which was continued for more than forty years. For this purpose a telescope of very long focal length was designed at Lick Observatory, and dragged from the site of one total eclipse to the next. It was 12m long and on its outsize plates produced a picture of the sun's disk of about 10cm in diameter.

This systematic activity, direct observation of the corona during total eclipse, spectrum analysis, and other methods gradually built up a picture of the state of the corona and of the protuberances. composition, density, temperature, and relation between form and appearance of the corona on the one hand and the magnetic events on the sun on the other were established. But contrary to expectations it was not possible to detect changes in the corona during a total eclipse. Several attempts to construct a coronagraph failed until Lyot's successful design in 1930. In the clear air of Pic du Midi at an altitude of 2,680m he was able to deflect the stray light and to produce and photograph a picture of the corona. Eventually, after several failures, he succeeded in observing, examining, and photographing the spectrum of the corona, although this was confined to the inner corona.

Today it is possible to follow the corona with the coronagraph out along about two solar radii from the sun. But the photographs obtained during total eclipses reveal the corona to distances of five or six solar radii from the sun. This difference is due to the scatter of sunlight in the earth's atmosphere by dust particles, which render detailed observation impossible even from the highest mountains. Even as late as the 1940s and 1950s we still depended on total eclipses for the examination of the outer corona. This situation changed only with the advent of rocket and satellite technology.

To begin with, coronagraphs were carried on high-flying aircraft and manned balloons; later, on high-altitude rockets and artificial satellites. Rockets and satellites naturally called for considerable electronic installations because the data had to be transmitted by the satellites, but this difficulty has been overcome through further progress in electronics.

After a search lasting over a century, it became possible for the first time in 1971 with the OSO7 satellite to follow the ejection of coronal material into interplanetary space in the light of the white corona. But the real breakthrough came with Skylab. Not only was Skylab manned by astronauts, who are able to adapt themselves and their conditions of observation to changing situations; but Skylab had enough space to accommodate complex instruments capable of producing an optimum yield of information.

This space station offered the additional advantage of conducting solar research in absolutely ideal conditions. At its orbit altitude of 435km the exploration of all solar phenomena was free from the restrictions imposed by the atmosphere. Every type of radiation reaches this region; absorption or reflection back into space of radiation such as the long, medium, and short radio waves and the infrared, ultraviolet, and x-ray waves begins at a level much lower than 435km.

Almost as difficult as the investigation of the corona was investigation of the protuberances—reddish, flamelike structures rising from the surface of the sun. Sometimes the protuberances appear stationary, but at other times they are ejected upward by tremendous explosions. Before the development of the spectroscope, protuberances could be seen only during total eclipses.

The smallest protuberances are about the same size as the earth;

the largest rise 1 million kilometers into space. Those extending to 150,000km above the surface of the sun are quite common. They, too, follow the magnetic field lines and are shaped by the interplay of forces between magnetism and matter. In spectroheliograms they often appear projected onto the surface of the sun as dark, threadlike structures called filaments. The spectrum of these phenomena largely corresponds to that of the chromosphere; we now believe protuberances consist of cooler and denser matter which penetrates the hotter and thinner corona. The shape of protuberances varies widely. Some shoot explosively into space at speeds of up to 100km/sec and completely detach themselves from

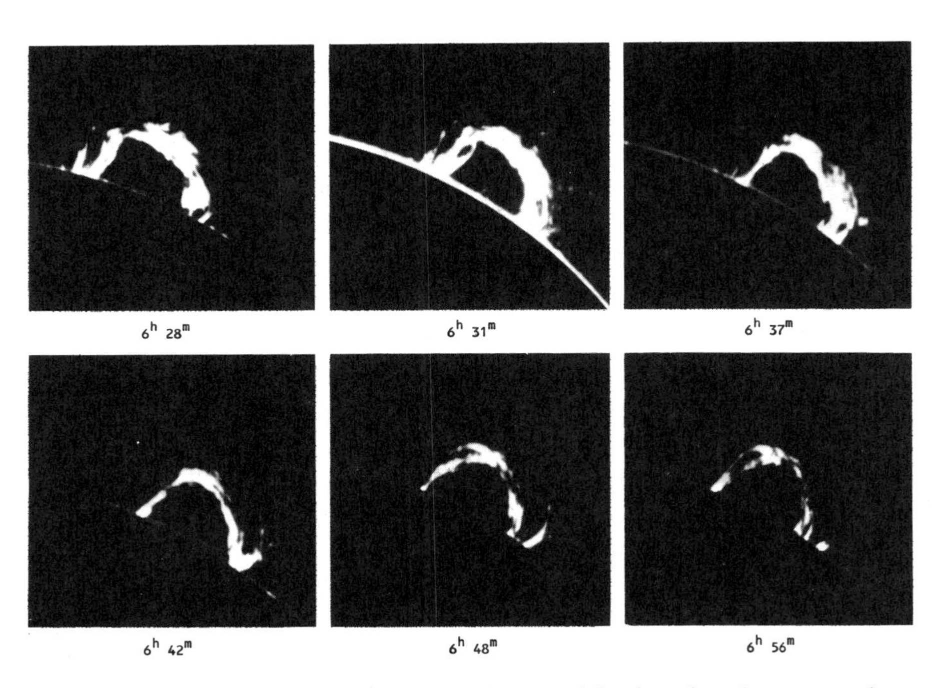

This series of photographs, taken at the Wendelstein solar observatory in the summer of 1948, shows how quickly eruptive protuberances change. The pictures were exposed within a period of twenty-eight minutes. Note the changes in the fine structure (typical of active protuberances of this kind).

the sun. Others occur in the form of arches, loops, and semicircles, rising only to rain immediately down onto the sun. Others remain motionless above the surface of the sun, hardly changing for days, weeks, and months. A connection often exists between protuberances and sunspots, when matter flows from protuberance to sunspot.

Extent, activity, and frequency of the protuberances rank equally with the sunspot cycle, but their maximum usually occurs before that of the sunspots. They extend into higher latitudes of the sun than the sunspot foci. Sunspots are limited to certain heliographic latitudes and in their timing depend on the phase of the given sunspot cycle. The same applies to protuberances, although their dependence on the phase of a given sunspot cycle differs from that of the sunspots on the heliographic latitude. The shape, appearance, and occurrence of protuberances are closely connected with the local magnetic fields on the sun's surface. Protuberances often "trace" the magnetic field lines. Those that suddenly explode and enter interplanetary space affect the upper atmosphere of the earth.

From Skylab the protuberances were explored in various regions of the spectrum, mainly in ultraviolet light. This produced stratified profiles of the structures and temperatures of these phenomena, which enabled us for the first time to understand the structure of the protuberances and to follow their development. Layers of progressively hotter matter surround cooler cores; the Skylab astronauts were able to trace them right into the corona. It was shown during this exercise that the protuberances are much more extensive than they appear to be in visible light; only in ultraviolet light did the hottest zones register.

The Skylab astronauts were the first to observe the *entire* atmosphere of the sun and to trace magnetic field lines within the corona. They obtained the first pictures that clearly revealed the connection between the events on the surface of the sun and the shape of the outer corona. The details of the reactions that take place in the transition zone between the cooler chromosphere and the hot corona were also successfully investigated. On photographs of the sun in the ultraviolet and x-ray regions of the spectrum the astronauts found "corona holes," empty spaces without any hot gases, but where cooler coronal gas leaves the sun in a stream of particles.

This is solar wind which we meet throughout the planetary system except where it is deflected by planets' magnetic fields. The impact of the solar wind affects the magnetic fields of the planets; it compresses them frontally. Such effects can be observed in the magnetic field and the upper atmosphere of the earth. The discovery of the source of the solar wind in the corona has enabled us to improve forecasts of the effects of solar wind on interplanetary space and planets—especially the earth.

Spectrum analysis has revealed temperatures in the corona of 1 to 2 million degrees C. In some local areas of the corona, at altitudes from 10,000 to 100,000km above the photosphere, a temperature as high as 4 million degrees C has been measured. It is not yet possible to offer a satisfactory explanation of the high temperatures of the thin coronal gas. They obviously are connected with the photosphere. The bubbling granules there—balls of gas several hundred kilometers in diameter and rising at several times the speed of sound—generate shock waves that continue into the corona, where they heat the thin masses of gas through impact.

The corona, the source of a large part of the x-ray and radio emis-

A solar protuberance in various temperature ranges (July 10, 1973).

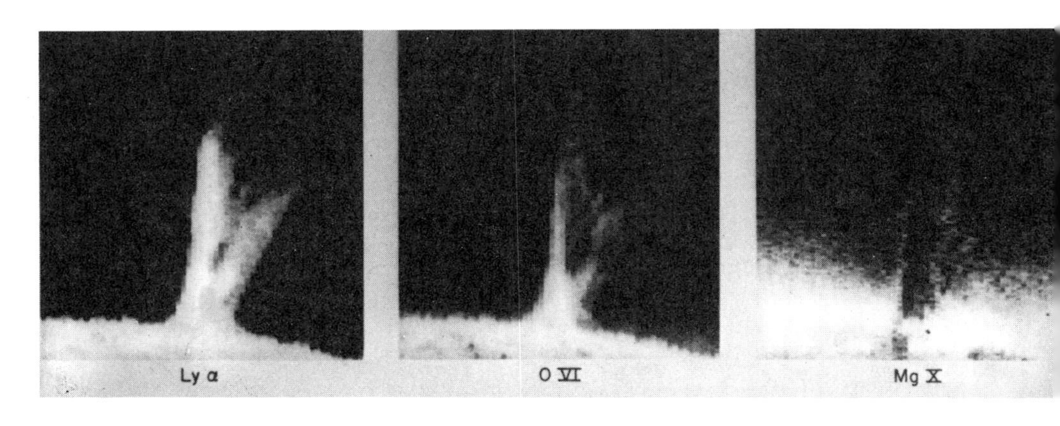

Left: in the light of the Lyman-Alpha line in the ultraviolet region of the spectrum at 10,000°C. *Center:* in the light of 5× ionized oxygen at 50,000°C. *Right:* in the light of 9× ionized magnesium in the corona at temperatures of several million degrees.

sion of the sun, extends far into interplanetary space. It is a direct link between the sun and the interplanetary matter that triggers the zodiacal light. This is a pyramid-shaped phenomenon that occurs in twilight at certain times—immediately before sunrise in the spring and just after sunset in the fall—and is caused by the scatter of sunrays by dust particles and electrons that fill the zodiacal light region in the vicinity of the sun.

THE CHEMICAL COMPOSITION OF THE SUN

Spectrum analysis has given us the first inkling of the chemical composition of the gases on the surface of the sun and in the layers of the chromosphere and the corona. The sun's surface consists of about 90 percent hydrogen, 8 percent helium, and 2 percent heavier chemical elements. In the interior of the sun, however, this ratio changes in favor of helium. As we have seen, the sun obtains its energy from processes of nuclear transformation from hydrogen to helium. As a result, more and more hydrogen disappears and fresh helium is produced over time. It has been possible to calculate the proportions. The sun produces 650 million tons of helium every second. At the same time it loses 655 million tons of hydrogen through transformation. According to Einstein's mass-energy equivalence the difference is converted into radiation and heat. The loss of solar mass thus amounts to 5 million tons per second as a result of energy production. At first glance this appears very high, yet it assures the sun of a life lasting many billions of years. At present, as a consequence of nuclear transformation, the sun as a whole consists of about 75 percent hydrogen, 23 percent helium, and 2 percent heavier elements. In the course of eons this composition will continue to change quantitatively from hydrogen to helium.

Besides hydrogen and helium, most elements are represented in the sun in varying quantities but not adding up to more than 2 percent of the sun's total mass. Of the ninety-two stable elements known to us, fifty-seven have been identified in the sun. For seventeen of the remaining thirty-five, some unreliable evidence has come to light; no evidence exists for the last eighteen. A number of these elements, however, can hardly be demonstrated even if they do exist because all their spectrum lines are confined to regions of the spectrum that are difficult to reach. Some rare earths, several

The corona of the sun in white light, photographed with the coronagraph on Skylab on June 10, 1973. The picture shows an eruptive protuberance, 220 times the diameter of the earth, expanding at a velocity of 1.5 million kilometers per hour (more than 400km/sec). The tenuous outer corona cannot be observed from earth.

noble gases, halogens, and all elements heavier than lead have not been detected so far. The following elements are heavier than helium and are common in the sun. They are listed in the order of their proportions: carbon, nitrogen, oxygen, neon, sodium, magnesium, aluminum, silicon, sulfur, argon, calcium, iron, nickel.

Molecular compounds have been identified in the sun. Some—such as the CN (cyan) molecule (a compound of an atom of nitrogen with one of carbon), the CH (hydrogen-carbon) compound, and various other hydrogen compounds—do not exist on earth. About 40 percent of all the Fraunhofer Lines in the solar spectrum have not yet been identified. They are mainly faint lines in regions containing a large number of spectrum lines. The most frequent of them are those of the iron group. Iron alone accounts for 3,300 lines in the solar spectrum; titanium and chromium, for more than 1,000; cobalt, for almost 800; nickel and vanadium are represented by more than 600; and manganese by almost 500 lines. The spectra of the chromosphere and of the corona also include the numerous lines of elements that have undergone multiple ionization.

In this chapter the complexity of our sun has been described. However, we are still in the dark about many of its characteristic features. The sun and the processes that take place on and in it rule the life of every one of us.

It is fortunate that our sun is such a quiet star. It is not—at least at the present—inclined to explosions that would engulf the earth and its inhabitants in a raging fire. It is fortunate, too, that the sun does not yet suffer from a lack of nuclear fuel that could lead to a collapse of the thermal equilibrium—which would also mean the end of the earth. However, we shall not escape this fate. But that time lies so far in the future—several billion years ahead—that life on earth may long have ended for other reasons.

Other Suns

THE NEW ASTRONOMY OF THE FIXED STARS

On a clear, moonless night, far away from the disturbing light of cities, from 2,500 to 3,000 stars can be seen with the naked eye. They are all radiant suns like our own. Many are smaller, many larger, some older, others younger. What accounts for the preeminent position of our sun is its comparative proximity to us. Our sun is the only star close enough for us to be able to observe details. This allows us to gain insight into the mechanism of a star and has helped us to understand, at least to some extent, the workings of the other stars.

The history of the exploration of the stars was no less beset with difficulties than that of the exploration of the planets and our sun. As should be expected, there are great differences between the apparent aspect of the stars afforded by a brief glance at the sky and the reality hiding behind it. As with the sun and the planets, thousands of years of the most painstaking observation and recording were necessary before we could succeed in drawing a reasonably true picture.

The principle of observation and experiment had to become firmly established before it was possible to develop reasonably accurate ideas. We have now entered the fourth phase of development, the phase in which we send measuring instruments into space. But even the nearest stars are too far away for us to be able to send our space probes—now and in the foreseeable future. Space travel and space research nevertheless have initiated a change in stellar astronomy similar to that in the astronomy of our planetary system, where manned spacecraft, and certainly space probes, can reach some of the objects of our interest.

Again, the reason for this is the filter action of the earth's atmosphere, which prevents many types of radiation from reaching ground level and falsifies others almost beyond recognition. Thanks to the possibilities afforded by space travel, we can now explore the stars from outside the earth's atmosphere and thereby not only cover the regions of the spectrum inaccessible at ground level but also look much farther into the depths of the universe. Radio astronomy has also opened a new window on space and especially on the stars. This has led to the revision of many traditional ideas and brought to light many new facts. The astronomy of pulsars and quasars, of x-ray stars and black holes, of neutron stars and radio stars are branch astronomy that have developed only within the last few decades and to which research with the aid of balloons and artificial satellites has made a major contribution.

The new astronomy has made considerable strides and now enables us to outline at least a rough picture not only of the world of stars as we see it today, but of its past and, to a certain extent, its future.

The stars are too far away from us to reveal details of their surfaces, not even in our largest, most powerful telescopes. But we can obtain information about their size, surface temperature, chemical composition, and radial motion—toward or away from us—by means of physical methods similar to those employed in the exploration of the sun and planets. We can therefore group the stars in certain age classes, distinguish between young and old stars, between short-lived and long-lived ones. The great variety thus becomes a sequence in time, a picture of the life of the stars.

HOW TO DISTINGUISH BETWEEN STARS

When we raise our eyes to the night sky the first thing we notice is that the stars vary in brightness. We see a few very bright ones and many faint ones. Since Hipparchos (about 190–125 B.C.) stars have been divided into six classes of magnitude according to brightness in the sky. This scale—although extended and refined since—is still in use today. Hipparchos placed the brightest stars in the 1st, and the faintest stars, which could only just be discerned with the naked eye, in the 6th class of magnitude. After the invention of the telescope this scale obviously had to be extended because now stars almost without number were added, some of them so faint that they are invisible to the naked eye. Their light must be concentrated by the collecting power of the lens for our eyes to be able to see them. Thus the 7th, 8th, 9th, and later even higher magnitudes were added to this scale of stars. With the most modern telescopes today we can photograph stars of the 23rd and 24th magnitude, exposing the film for an hour and even longer.

Calibration of this ranking order of stars also revealed stars that are brighter than 1st magnitude. According to the principle of the thermometer negative magnitudes were therefore introduced. The 1st magnitude is now followed by 0 and −1st magnitude. A superior m (^{m}) for magnitude indicates that the brightness value refers to stars, not planets. The difference between two classes is now defined as 2.512; this means that a star of, for instance, 1st magnitude is 2.512 times as bright as a star of 2nd magnitude. A 1st magnitude star is accordingly 100 times as bright as one of 6th magnitude, and 1.6 billion times as bright as one of 24th magnitude. The brightest star in the sky, Sirius (in Canis major), has a brightness of -1.47^{m}; the full moon, −12.55; the sun, -27.74^{m}. We are now capable of determining the brightness of a star to an accuracy of hundredths of a magnitude.

The brightness of stars gives no indication of their relative distances; after all, one may be a very intense light source far away; another, a faint object very close to us. To establish comparisons absolute brightness (represented by ^{M}) has been defined as that of stars as if they were all at a distance of 32.633 light years.

The distances of the stars compared with those of the planets and our sun are almost beyond imagination. They are therefore measured in light years. The concept of light time was introduced earlier in this book. A light year is the distance a light ray travels within a year at a velocity of about 300,000km per second (the precise velocity of light in a vacuum is 299,792.458km per second). This is 9.460s $\times$ 10^{12}, or 9.4605 trillion kilometers—so, roughly, 10 trillion kilometers. We can visualize the enormity of this distance if we again compare the traveling time of an express train or an airliner. Our express train moving 60mph would require a little less than 11 million years, the jet liner more than 1 million years, to travel this distance and this is only a single light year. There is not one star that close to earth; the nearest one, a tiny star in the constellation of Centaurus (visible only in the Southern sky), is 4.3 light years away—46 million years on the express train or 4.6 million years on the plane.

The "awkward" number of 32.633 light years was chosen as the standard distance for the calculation of absolute magnitude because it is the distance of 10 parsec. Parsec is the contraction of the term "parallax second" and 1 parsec is the distance at which a star seen from the earth shows an annual parallax of 1 arc second or appears at a distance of 1 arc second seen from the earth and the sun. Expressed in light time this represents a distance of 3.2633 light years; 10 parsec = 32.633 light years.

What human imagination has composed into constellations from early time are stars that are basically unrelated; however, seen from the earth, they are all roughly in the same direction. All stars that we can see belong to the same huge cloud of stars, an island in space, a conglomeration of about 100 billion stars that we call our "galactic system." The members of one and the same constellation, however, are random objects, recognizable only from our solar system as constituting this constellation.

To place all these stars mentally at a standard distance and to indicate the brightness at which they would appear from there is possible as soon as we can determine the distance of a star or its true luminosity by means of geometrical, physical, or other methods. Although we cannot do this for every star, we can do it directly for those nearest to us, and for many stars we can resort to indirect or at least to statistical methods. The mean distance between stars in our own vicinity is 7 light years. But our sun is not in the center of our galactic system; that center lies in the direction of the constellation Sagittarius, about 30,000 light years away. The stars there are much closer together. The entire galactic system has the shape of a lens. Its major diameter is about 100,000 light years; its thickness in the central region about 16,000 light years, tapering off toward the edges.

When apparent brightness of stars whose distance or luminosity is known is converted into absolute brightness it becomes evident that some stars radiate considerably more intensely than our own sun and that others are very much fainter. If our own sun were at that standard distance of 32.6 light years, it would be a star of only 4.8 magnitude, but still clearly visible to the naked eye. Sirius (at

-1.47^{M} the brightest star in the sky) surpasses the sun's absolute luminosity by a factor of 22; its absolute brightness is $+1.4^{M}$. Sirius is 8.7 light years away from us, and therefore appears much brighter than it would appear from the standard distance of 32.6 light years. Rigel, the star representing the right foot of Orion, has an apparent brightness of $+0.3^{M}$ even though it is 1,300 light years distant from us; this suggests a very great luminosity or absolute brightness. Indeed, its absolute magnitude is -6.3^{M}. This means that the luminosity of this star is 422 times that of our own sun.

Some stars emit a much weaker radiation than our sun, at absolute brightnesses of $+10^{M}$; $+12^{M}$, or $+14^{M}$. But the apparent brightness of most of them is also not very great. A star of absolute brightness $+14^{M}$, for instance, has, after all, only about 1/260th the sun's luminosity.

A more detailed look at the dark, moonless sky, reveals stars of different brightnesses, and also of different colors. Some appear bluish-white, pure white, yellowish; some even appear reddish to red. For instance, Sirius and Vega (in Lyra) are bright yellow stars; Capella (in Auriga) and our sun are examples of deep yellow stars. Arcturus (in Boötes), like Aldebaran, (the main star of Taurus) is reddish yellow. Betelgeuse (the left shoulder star of Orion) and Antares (in Scorpio) are typical red stars.

Color differences indicate different surface temperatures. The blue stars have the highest, the red ones the lowest surface temperatures. A connection also exists between the color or temperature of a star and its spectral class.

Spectral classes cover groups of stars whose spectra share characteristic features. They have been arranged in sequence by color or surface temperature: the first (or "early") spectral classes include the blue and white stars of the highest surface temperatures; the last (or "late") classes are red stars. The terms "early" and "late," incidentally, do not suggest a chronological sequence of these classes at all; the chronological stages in the life of a star progress, as we shall see, in a completely different form.

Originally the spectral classes were divided according to the letters of the alphabet. A more accurate analysis of the classes, however, has not only disturbed this sequence, but also led to the abolition of some of the groups. Today, the sequence (starting with the hottest stars) runs O, B, A, F, G, K, M, R, N, S. A well-known mnemonic is "Oh, Be a Fine Girl; Kiss Me Right Now Sweetheart";

this may help you to memorize the sequence. The R, N, and S groups are "subgroups," especially carbon-containing and cyan- and zirconium oxide-containing cool stars; and the very hot O stars, at the beginning of the sequence, are very rare. But the spectral subdivision goes even further: the numbers 0 to 9 are added to the letter of the spectral class for more detailed distinction, hence, an A0 star is a pure A star, and an A5 star is halfway toward the F type. There are further identifying letters to stress other properties of the stars. For instance g represents a giant star; d is a dwarf star.

KEY TO THE TYPES OF STARS

Several parameters of the stars have been explained on the preceding pages; absolute brightness, absolute magnitude, luminosity, color, surface temperature. To link these and other parameters a graph can be plotted that includes some or all of these values.

In 1905 the Danish astronomer Ejnar Hertzsprung made some attempts in this direction; in 1913 the American Henry Norris Russell developed a comprehensive diagram which has come to be known as the Hertzsprung-Russell Diagram. It shows the spectral classes of the stars (and, thereby, their surface temperatures or colors) compared to absolute brightness or luminosity. The absolute brightnesses are plotted from top to bottom, and the spectral classes from left to right. If a representative number of stars is entered on this diagram astonishing relations will be revealed. It is not, as we might expect, that the same number of stars will be found throughout the diagram, that there are very bright stars of all spectral classes as well as very faint ones in all colors. What will be found is a relation between spectral class and absolute luminosity which sometimes permits conclusions about the absolute brightness of a star from its spectral class.

On the vertical axis absolute brightness is entered from top to bottom, with the brightest absolute magnitudes at the top and the faintest ones at the bottom; spectral classes, on the horizontal, beginning on the left with white stars and ending on the right with red stars, are entered on the horizontal axis. Most stars will be situated on a line running in the graph from top left to bottom right. This oblique branch is called the Main Sequence, with the luminous, hot stars at the top left and the fainter red, cool ones at the

The Hertzsprung-Russell Diagram.

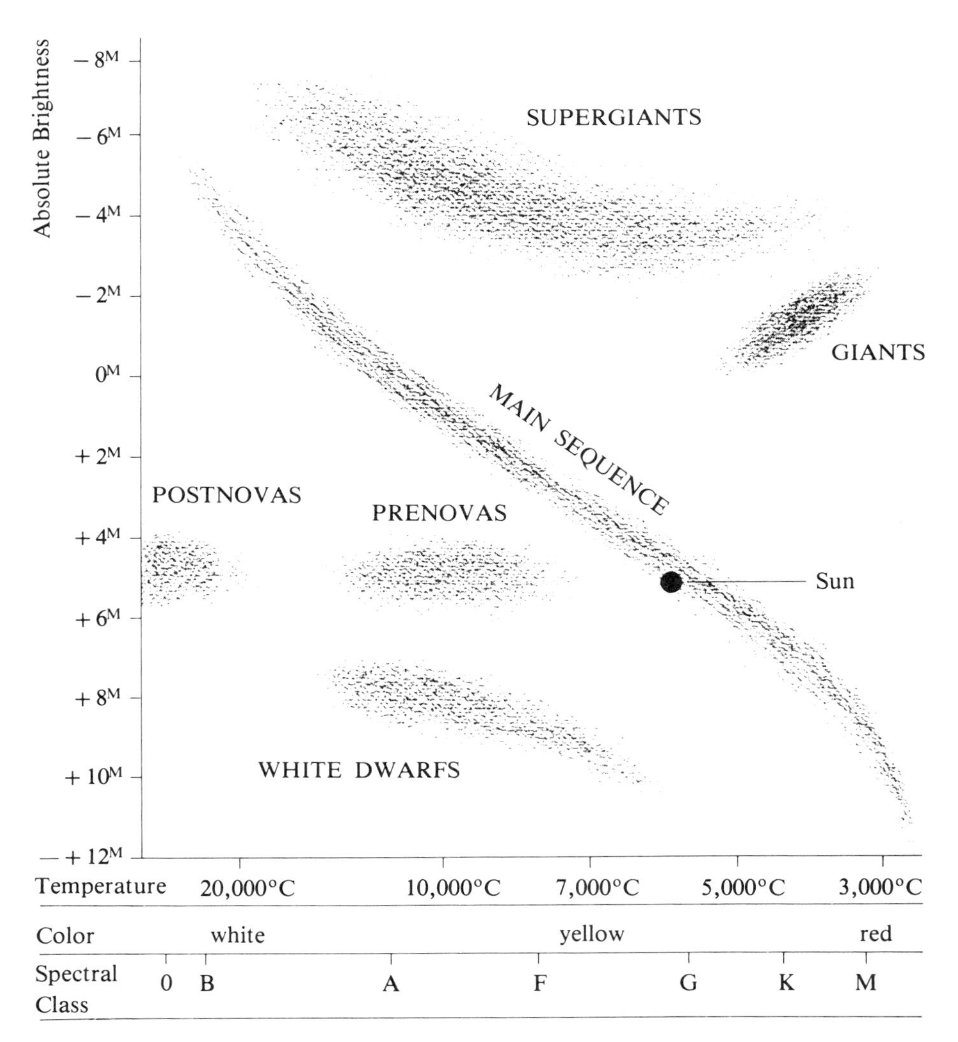

The diagram illustrates the relation between brightness or luminosity of the stars on the one hand and their spectral class and surface temperature on the other. It shows the Main Sequence which includes by far the most stars; the Giant Branch, consisting of "supergiants" and "giant stars"; the White Dwarfs; and the stages of novas before energy eruption (prenovas) and after it (postnovas). In the past it was believed that a star began life as a Red Giant, moved into the Main Sequence, and, after passing along the sequence, would slowly shrink and end as a Red Dwarf. We now know that the evolution of stars is more complicated.

bottom right. A second branch covers very luminous yellow and red stars, which are at the bottom right end of the diagram. This part of the Hertzsprung-Russell Diagram is called the Giant Branch. We also find a few stars of small absolute size but emitting white light on the bottom left of the diagram; they are called the "White Dwarfs."

Only more detailed statistics, confined to the stars of a certain vicinity, provide information about the frequency of the various types of star. The Red Dwarfs on the bottom right branch of the Main Sequence are, accordingly, the most common; there are comparatively few Red Giants, outnumbered even by the White Dwarfs. Generally faint stars of weaker luminosity than the sun are more common than luminous ones—at any rate in our part of the universe. There are about 100 times as many stars of absolute brightness from $+8^{M}$ to $+12^{M}$ as those of $+5^{M}$, the absolute brightness of the sun.

Mass and Luminosity

The lives of the stars vary with their types. At one extreme are stars whose luminosity is only one ten thousandth that of the sun; others are a hundred thousand, indeed a million times as luminous. The former are dwarfs in every respect; the latter, giants not only in luminosity but in energy consumption, age, and dimension. But there are also stars of different luminosity, different energy emission per unit area. The mass of a star, the quantity of matter it consists of, is also important. Some stars have the same mass as the sun, others more than a hundred times; still others have only 0.04 times its mass. The overwhelming majority of stars have masses ranging between 0.3 and 3 solar masses.

Clearcut relations also exist between stellar mass and luminosity. They were represented in a diagram by the British astrophysicist Sir Arthur Eddington in 1924 for the stars of the Main Sequence and expressed in a formula. According to Eddington's formula the luminosity increases in proportion with the 3.5th power of the mass: $L = \text{const } M^{3.5}$; expressed in logarithms: $\lg L = \text{const.} + 3.5 \lg M$, where const = a certain constant.

However, this equation covers only Main Sequence stars; the Red Giants and the White Dwarfs deviate greatly from this relation. For the Main Sequence at any rate this mass-luminosity relation is accurate enough to be used for the calculation of a star's mass from its known luminosity.

The mass of a star has shown itself to be a very important factor in its life. The bright, massive stars of the Main Sequence are absolutely profligate; they yield the brightness and heat derived from their nuclear energy so quickly that some of them, Supergiants such as Rigel (in Orion) in the extreme top left of the Hertzsprung-Russell Diagram, attain life spans of only a few million years; the fainter, less massive stars can survive for up to 10 billion years before the inevitable instability and collapse.

THE LIFE STORY OF THE STARS

Let us look at the birth of a star before we consider its death. Even with the naked eye a small patch of nebula can be seen below the three stars of Orion's Belt. This Sword Belt becomes more prominent through binoculars or through a small telescope. This nebular structure is about 1,600 light years away from earth; if we consider only the inner portions of the nebula its diameter is about 15 light years. If we include the thinner outer portions, the diameter of the Orion Nebula is 50, perhaps even 100 light years, and its total matter is about 700 times that of the sun.

This cloud of dust and gas is a typical birthplace of stars. Here we have a large conglomeration of O stars, the hot, bluish-white stars. But sources of intense infrared radiation, too, were detected in the Orion Nebula within the last few years.

According to current views stars originate in interstellar matter, cosmic gas, and cosmic dust. Whereas the "empty" interstellar space at a density of 10^{-24} g/cc contains on average a single hydrogen atom per cc (atmosphere at ground level contains almost 10^{-20} atoms per cc), we find between 100 and 1,000 atoms per cc in the numerous gas and dust clouds such as the Orion Nebula, and as many as 10,000 atoms per cc in the densest central regions.

How such condensations of gas and dust atoms come about has not yet been fully established. According to one theory condensations are triggered by waves of matter sweeping through the galactic system and causing local condensation attenuation of the extremely thin material. Another view claims that they are shock waves, supersonic waves that travel faster than sound and are generated during the explosions of supernovas. Here, too, local condensation of gas and dust is produced, which results in protostars.

A third opinion states that stars are born in conjunction with hot stars already existing in the bright nebulae such as the Orion Nebula. These stars, called O stars, have surface temperatures from 50,000 to 100,000°C and heat the hydrogen in the gas clouds surrounding them to about 10,000°C; the hydrogen becomes ionized in the process. But the ionized hydrogen spreads into the outlying neutral hydrogen at an absolute temperature of only 100°K (−173°C). (Absolute temperatures, measured in Kelvin degrees, have their zero point at −273°C.) At the interfaces between the

Infrared Radiation

Infrared is a region of the spectrum of wavelengths of about 8,000 A.U. (1 A.U. = 10^{-8} or $\frac{1}{100\ 000\ 000}$ cm). This wavelength is too long for the eye to register. Since about 1965 astronomical research has been conducted in this region, mainly from balloons sent up to altitudes of 30, 40, and 50km. This distance is necessary because water vapor and carbon dioxide in the atmosphere absorb most infrared wavelengths. Some of the infrared radiation of celestial objects cannot be detected at ground level; other parts of it only on top of high mountains.

Emitters of infrared radiation in the universe were detected during the latter half of the 1960s, first at the Mount Wilson Observatory, and later with the aid of balloons. It was mostly the infrared radiation of cool stars with surface temperatures no higher than 1,000 to 2,000°C. The infrared radiating objects in the Orion Nebula, on the other hand, are hot stars whose radiation is largely absorbed in the shorter wave ranges—invisible light—by the surrounding interstellar dust clouds, so that only the infrared radiation escapes. Infrared nebulae, too, were found in the Orion Nebula. These are as yet unborn stars still in the "embryonic" stage of their development. The whole Orion Nebula is probably only a few thousand years old. Some of the stars in it are younger still, and the sources of infrared radiation are stars at the stage of being born.

neutral and the ionized hydrogen, condensed wave-shaped salients are produced; these separate from the cooler neutral hydrogen gas, and the neutral hydrogen condensed in them is compressed even more by the hotter ionized hydrogen of the surroundings. It then develops into a star, or, in the course of further divisions, into several stars or possibly into a whole star cluster.

There are globular, dark areas in the sky with diameters of about 150 billion to 15 trillion kilometers. These globules could be stars being born. The more the gas of such a star condenses, the more its temperature rises. When it attains 100 to 1,000°C the star begins to radiate infrared light—exactly like the infrared nebulae and the infrared stars that have been found in the Orion Nebula and in other comparable objects. With further condensation owing to gravita-

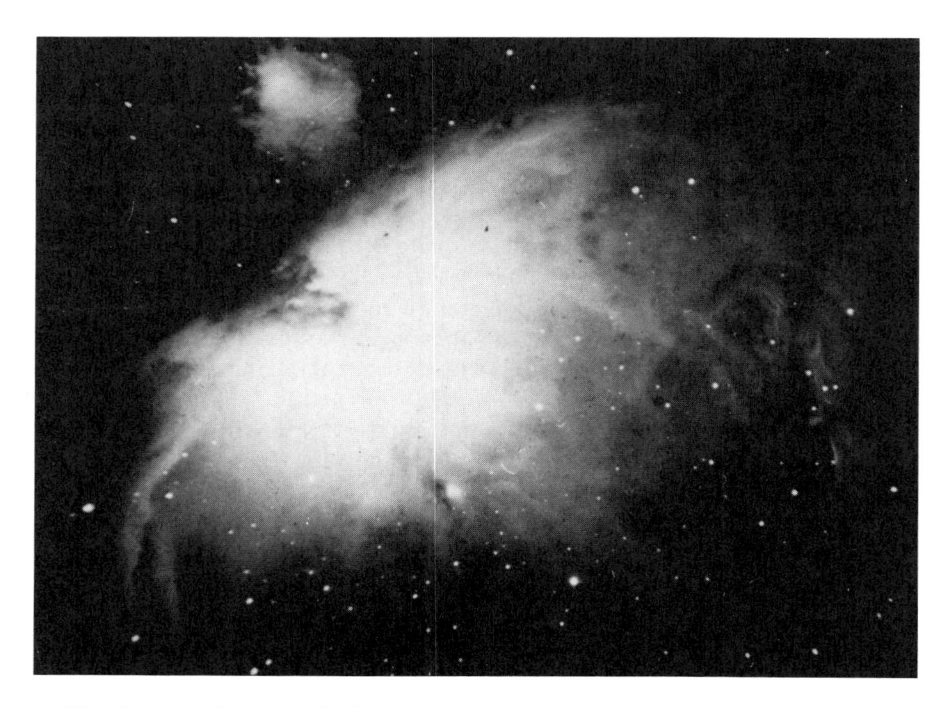

The famous Orion Nebula (M42 or NGC 1976) could conceivably be a center of star formation. It is a bright, diffuse nebula in Orion's Belt, barely visible to the naked eye. It is 1,600 light years away; the diameter of its center is about 15 light years (50 to 100 light years if the most tenuous marginal zones are included). The mass of gas of which it consists is about 700 times that of the sun.

This picture of the constellation Sagittarius, taken with the Schmidt reflector at the Wendelstein observatory, shows star clusters, gas nebulae, and dark nebulae. Schmidt reflectors are special telescopes for celestial photography; they have a particularly large field of view.

tional force (each of the particles of matter attracts any other particle, which leads to a pronounced concentration of matter in the direction of the center of the star, counteracted only by the increasing internal pressure of the star) the temperature increases until ultimately it becomes high enough for nuclear fusion processes to start. This is the real hour of a star's birth.

Estimates have been made of how long it would take a star to develop in this way. The size of the protostar and the quantity of the available material are obviously important factors. Stars of very great mass could sufficiently condense even within 10,000 years for nuclear fusion and thereby a liberation of nuclear energy to set in. Smaller stars, like our sun, require 30 million years; and stars of 1/10th or 1/100th the mass of the sun, about 100 million years.

We have learned that hot, luminous stars waste energy. The life of a star begins as a protostar in the top right part of the Hertzsprung-Russell Diagram. From there the star moves to the left to the Main Sequence of the stars. There it will maintain a place during the greater part of its life span, and this place depends mainly on its mass. Massive star bodies develop into very hot, bluish-white stars of enormous luminosity and high surface temperature; on the other hand, stars of small mass end up somewhere at the bottom right end of the Main Sequence as Red Dwarfs. All the other stars are in the intermediate region.

But no "order of rank" of the stars must be deduced from these considerations. The hot, bright, massive stars must not be regarded as vigorous specimens, and the fainter, less massive, cooler ones as inferior ones. As so often in life, stars, too, have to pay for advantages they enjoy with disadvantages. The hot, luminous stars pay for their brilliance with a shorter life. They are bankrupts whose fate is already sealed at the hour of their birth. Dwarf stars and yellow stars (like the sun) remain in the Main Sequence of the Hertzsprung-Russell Diagram for many billions of years; the white and blue giants, in spite of their huge masses, may drift out of the Main Sequence after 10 million years. This also explains why we find a relatively small number of supergiants. During the history of the universe there was bound to be many supergiants, but at a certain period there are comparatively few of them, because so many have already reached the end of their life and have disappeared from the Main Sequence of the Hertzsprung-Russell Diagram.

A star enters the Main Sequence of the Hertzsprung-Russell Diagram when atomic fire begins to burn in its interior. Until then the star obtained its luminosity from the contraction energy alone, from the progressive concentration of matter to increasing density as a result of gravity. For nuclear fusion to start, temperatures of the order of several million degrees C and densities of 10 to 100g/cc are necessary. The decisive process of fusion, incidentally, is the same as that in the sun, the transformation of hydrogen into helium. There are other types of nuclear fusion, in which a few light elements are produced—heavy hydrogen, lithium, beryllium, boron—but they are relatively unimportant to the energy balance of the star.

During the phase at which energy gain from nuclear fusion starts, the contraction from which the star had obtained its energy up to that point comes to an end. The surface temperature now rises sufficiently for the star to begin to radiate within the visible region of the spectrum. It has become one of the many familiar stars in the sky. Its further development depends on the chemical transformations that take place inside it as a result of nuclear fusion.

We have learned that changes in the chemical composition of very massive stars caused by nuclear fusion proceed rapidly by cosmic time scales; sometimes a star will die after only a few million years. But other stars, such as our sun, pass through their life in many billions of years and some of them have so little mass that nuclear fusion never comes about; they are stars we could call stillborn.

A nebula in the constellation Unicorn, photographed in the red region of the spectrum with the Schmidt reflector at Mount Palomar.

THE MISSING NEUTRINOS FROM THE SUN

We now believe we are quite well informed about the development of the stars. It has been possible within the last few years to examine various star models with the aid of computers which reproduce the observable facts very accurately. But in doing this we must disregard the problem that exists in the context of the energy production of our sun, a problem that has lately induced some experts to ask whether we are perhaps not on a completely wrong track. It is the problem of the "missing neutrinos."

We have seen that with energy liberation by helium from hydrogen in the sun neutrinos—extremely elusive elementary particles—are repeatedly released during the various reactions. We have also learned that it will be many millions of years before energy liberated during nuclear fusion inside the sun reaches the sun's surface. This is not the case with neutrinos, which are formed in large numbers during nuclear fusion. On their way from the sun's core they pass through the intermediate layers at the velocity of light without major reaction with other elementary particles and quickly escape into space. They also pass through most celestial bodies they meet without leaving any trace. The sun, the earth, and other such objects are, "transparent" to neutrinos almost as a window is to light. As they travel through the sun only about every billionth, and through the earth every trillionth, neutrino is involved in reactions. It has been calculated that on average a neutrino covers a distance of 10,000 light years before it is absorbed by other particles.

About 7 percent of the energy liberated by the sun is emitted in the form of neutrinos. As a result every square centimeter of the earth's surface is pierced by 60 billion of these particles every second. If they could be detected on earth it would be direct proof of nuclear fusion processes in the interior of the sun.

To find these solar neutrinos, Raymond Davis, at Brookhaven National Laboratory, began in 1955 to set up a most unusual experimental plant. It utilizes a "neutrino-capture reaction" conceived in theory by another scientist as far back as 1946. Davis and his collaborators installed a huge tank in an old gold mine in South Dakota—1,500m below ground—and filled it with about 400,000 liters of perchloroethylene (C_2Cl_4), a cleaning fluid. The so-called activation cross section, a value that expresses the frequency of nuclear reactions, is quite high for the stable chlorine isotope 37 for neutrinos of certain energies. Thus it should be possible to stop some of the solar neutrinos arriving on the earth and to involve them in reactions with these chlorine atoms. If an atom of chlorine 37 is hit by a neutrino, it becomes an atom of the radioactive argon isotope 37 and emits an electron.

$$\nu + \mathrm{Cl}^{37}_{17} \surd\leftarrow \mathrm{Ar}^{37}_{18} + \mathrm{e}- .$$

At a half-life of thirty-seven days the argon atom decays into a chlorine atom by capturing an electron from its own atomic shell. Again, the half-life of a radioactive isotope is the time required by one half of the atoms of a given radioactive substance to decay into a different isotope. Every radioactive isotope has a constant characteristic half-life. The energy liberated can be measured with suitable instruments, and the number of decay processes per unit time can thus be determined.

In the many years this experiment has been running, attempts to detect neutrinos from the sun have failed. Only about one tenth of the postulated "counting events" of argon 37 atoms were recorded. They are probably caused largely by the cosmic radiation, which is the reason why the experiment has to be carried out in the depth of a gold mine, where most of the cosmic rays are absorbed by rock. But residual amounts of cosmic radiation must be considered and, on average, they agree with the demonstrated tenth of the neutrinos expected from the sun.

This result, repeated several times and checked in vain for errors in the instrument setup and interpretation of measurements has left the scientists rather speechless. Some desperate attempts at explanations were made—none satisfactory; none without inherent contradiction. The astronomers accused the physicists that the suggested capture cross sections derived experimentally from other investigations were incorrect. But the physicists rejected this accusation as indignantly as the astronomers spurned the physicists' claim that the pressure of temperature conditions inside the sun differ from those the astrophysicists suggest. If this were so the consequence would be a different sequence of nuclear reaction inside the sun—in such a way, for instance, that it would not be the proton-proton reaction, but the Bethe-Weizsaecker Cycle that played the main part. But this does not work either, for the latter cycle yields even higher values for neutrino production and would make the discrepancy larger instead of smaller. Nor have considerations whether the neutrinos enter, perhaps in the sun, or on the way to the earth, into other still unknown interactions produced any results.

Perhaps the most bizarre explanation is the suggestion that the sun is unstable and that, as a result, the temperature in its center dropped by 10 percent a million years ago. This would produce changes in the nuclear fusion reactions and a reduced neutrino flux. Here on earth we could not have noticed anything of this decreased energy production inside the sun other than through the reduction in the neutrino flux, because, as we have learned, the energy transport from the core of the sun in the form of electromagnetic radiation takes 10 to 50 million years. Others suspect that the sun has ceased its energy-producing nuclear fusion processes altogether and postulate the imminent demise of the sun. In theory this makes little sense for according to all other findings the sun must be accorded a life span of at least another 5 billion years; even so, this would be of no importance whatsoever to life on earth today and in the foreseeable future. Many millions of years would pass before the sun reached that unstable state.

The problem of missing neutrinos will obviously persist for the time being. Other methods for detection of solar neutrinos from various nuclear reactions have been postulated, but the appropriate experiments have not yet been begun. How astrophysicists will solve this problem cannot be foreseen. Perhaps only a minor correction is necessary to save the situation; on the other hand, it may be a fresh indication of changes in physics of gigantic proportions which could lead us to completely new ideas.

We must at any rate admit that we cannot speak of a full understanding of the nuclear energy production in our sun, and therefore in other stars, until this problem is solved. Further experiments are currently being prepared with a new detector isotope, which may solve the riddle in the next few years.

This emmission area in the constellation Cygnus is called the North America Nebula because of its shape. Its designation is NGC 7000 (NGC indicates a catalog of nebulae and star clusters compiled by J. Dreyer in 1888). The object can be photographed with smaller instruments, but is visible only through powerful telescopes. The mass of the North America Nebula has been estimated at about 3,000 times that of the sun, and its density at 10 atoms/cc.

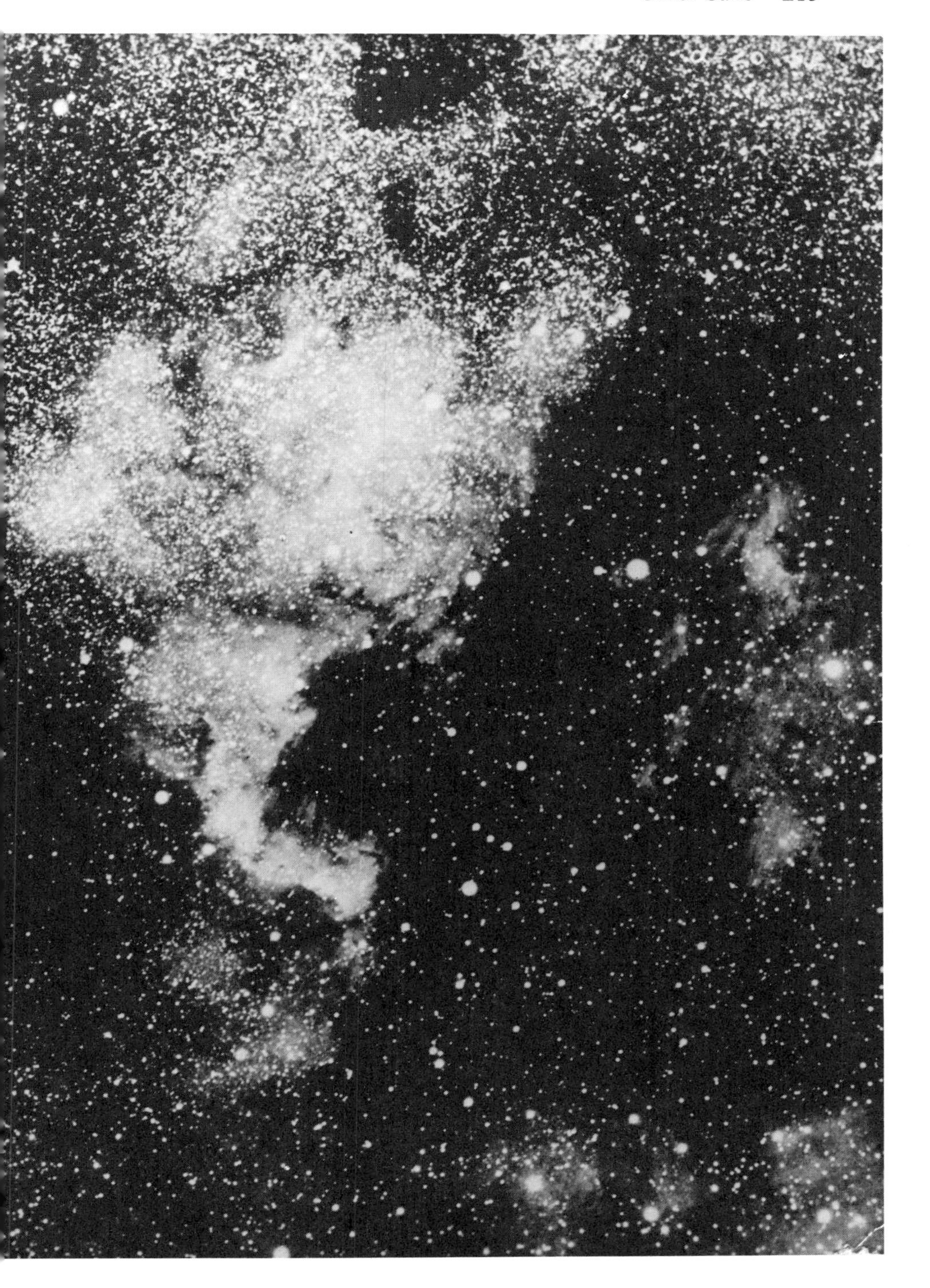

OLD STARS

What will happen when a star has reached the end of its nuclear hydrogen fusion, used up its nuclear fuel, and can no longer obtain energy from burning hydrogen and helium?

First, it must be pointed out that this situation will arise long before the very last reserve of hydrogen is exhausted; nuclear combustion will be choked off by the "slagging" of the star through the helium produced. This can be compared with a fire in a furnace that goes out when the proportion of slag (the coal already burnt) greatly exceeds that of the coal still available. In stars old age progresses according to a quite complicated scheme. It depends on a star's position on the Hertzsprung-Russell Diagram throughout its life—whether it was a dwarf that quite slowly burned for billions of years or a luminous, massive star that used up its capital within a few million years.

To begin with, let us take our sun as an example and see how other stars will fare in their old age. Our sun is a star of medium mass and spectral class G2, that is, a yellow dwarf. We expect it will remain such for another 2 to 3 billion years. By then the hydrogen in its core will have been gradually depleted through contamination by an excess of helium "slag." As a result the atomic fire—the transformation of hydrogen into helium—will be shifted to a zone around the initially small helium core that forms, a process of "shell burning."

Initially, this change will hardly be noticed on earth, but in the course of time a slight increase in the luminosity and surface temperature of the sun will be recorded. Slowly but steadily the sun will rise vertically on the Hertzsprung-Russell Diagram into the region of greater luminosity. At the next stage the surface temperature of the sun will reverse and become a little cooler than before. The sun's core will condense. This will produce another rise in temperature and luminosity. In just under 5 billion years the sun will be about twice as luminous and large as now; and the human race, if it still exists, will long have experienced great difficulties because of the enormous climatic changes that will have been wrought on earth as a consequence of the increasing radiation of the sun.

The more hydrogen burned inside the sun, the greater the central helium core will become and the farther out the hydrogen-burning zone in the sun will move. At the same time the pressure in the center will increase, because the helium nuclei of the "ash" are more compact than the hydrogen nuclei that had previously been predominant there. Thus the sun's core will collapse from the gravitational force. This will trigger higher temperatures which will spread through the entire sun and enormously dilate its outer regions. Its total energy production will then be greater than at present but, owing to the dilation of its outer layers, a much larger area will be available for radiation. As a result, the available quantity of energy per unit area will be reduced in spite of greater total radiation. The surface temperature will drop from its current value of just under 6,000°C to 4,000 or 3,000°C; the spectral radiation maximum will move into the long-wave region; the color will shift toward the red. The sun will turn into a Red Giant, a star of a diameter a hundred times and of a luminosity a thousand times that of the sun today. Mercury, Venus, and Earth will have become engulfed in an ocean of flames by this red sun long after their oceans evaporated, lead and rocks melted, leaving charred planets behind.

The matter in the interior of the sun will by now be so densely packed and so hot—with degenerate matter—that it no longer conforms to the normal relations between the pressure of a gas and its temperature according to which pressure is the product of density and temperature.

$$P \sim \varrho \times T$$

In this equation, P = the gas pressure; ϱ = the gas density; and T = the absolute temperature. But the inflating gaseous envelope of our sun will be considerably thinner than its atmosphere today.

Roughly during this phase—which will perhaps occur in 7 billion years—another important event will take place inside the old sun. The helium slag in the center of the sun will become a nuclear fuel, a new nuclear fire will be lit. The helium will burn to be transformed into carbon and into other heavy elements, as heavy as iron. This new fusion process starts within a few minutes or hours and is called helium flash. But the explosive ignition of the helium will not take place before a temperature of 100 to 200 million degrees C has been reached in the core of this Red Giant, and the density will have increased to the absolutely fantastic value of 1,000,000g/cc. A thimbleful of this degenerate stellar matter would weigh 500kg on earth.

The helium flash is a violent but brief episode during a star's old age, a fate that not all stars suffer. Only small, less massive stars like our sun are so affected. If a star is heavier than 1.4 solar masses, the helium flash will not occur.

This intense nuclear fire lasts for only 200 years—a fraction of an instant in the life of a star. Incidentally, the helium nuclear fire is not as energy-productive as the nuclear fusion of hydrogen into helium. The yield of these higher-order nuclear reactions is only about one twentieth that of the hydrogen-nucleus reaction. The elements that are produced range from carbon, the result of the fusion of three helium nuclei, through oxygen, neon, and in further processes from carbon through sodium, magnesium, silicon, and sulfur to iron. These processes also represent a major part of the story of how these elements evolved in the universe.

What happens when the nuclear combustion of helium to elements of higher atomic numbers has passed through its stormy episode? The nuclear transformations will continue, although the nuclear fire will burn more evenly because the energy yield is no longer very great. Disturbances will now occur also in the shells, which makes the further calculation of the star model impossible. Instabilities will take place at intervals of a few centuries that cannot be accommodated in the context of computations covering millions of years.

It appears, at any rate, to be certain that the Red Giant will, after the helium flash and a temporary decrease in brightness, once again enormously expand its outer shells to reach about 400 times its diameter and 10,000 times its brightness today, and to strip itself of its outer shells. Its core, on the other hand, will collapse further. The energy to produce these changes is gravitational; it is derived from the collapse of the star. The repulsion of matter basically occurs throughout the entire Red Giant stage; owing to the low gravitational acceleration in the outer layers of these "blown-up" stars matter is continuously emitted in the form of ejected masses of gas. This "stellar wind"—analogous to the solar wind—will be accompanied by gigantic protuberances. With the progressive contraction of the core this process will increase in scale. A number of objects we can observe in the sky—planetary nebulae—may be such stars in the throes of their last convulsions.

We know of about 1,000 planetary nebulae today, but it is assumed that there are about 50,000 of them in our galactic system. The name is misleading and has nothing to do with their cosmic classification. They were called planetary nebulae because the shape—a greenish disk—they present in the telescope closely resembles the appearance of Uranus and Neptune. They are, as a rule, remote stars that have reached the end of their life, and in their last convulsions emit parts of themselves as masses of gas, mainly hydrogen. The nearest, listed as NAC 7293 or Helex Nebula in the constellation Aquarius, is at a distance of 450 light years; the most remote planetary nebulae known to us are about 35,000 light years away. They tend to concentrate toward the center of the galaxy and are found in regions containing mostly older stars.

This agrees with the theory that the planetary nebulae are masses of gas ejected by dying stars. The gaseous envelopes of these objects expand, as we have been able to find out from measurements, at speeds from 20 to 50km/sec and sometimes even faster. A hot central star occupies the center of the planetary nebulae, and radiation excites the nebulae into luminosity. These central stars have a surface temperature of perhaps 50,000 to 100,000°C; they are smaller than our sun, but have roughly the same mass, and so are very dense.

The diameter of a typical planetary nebula is 40,000 astronomical units, or 6 trillion kilometers. Its average age is 10,000 years; presumably it lasts no longer than 50,000 years. The mass of the individual gaseous envelope probably ranges between 0.6 and 4 solar masses. Repeated strippings of outer shells by dying Red Giants could lead to considerable losses of mass by the original giant stars. They might drop below the limiting value of 1.4 solar masses, which would direct their lives toward a further final stage, that of the White Dwarf. The central stars of the planetary nebulae could be in a stage immediately preceding that of a White Dwarf.

White Dwarfs are found in the bottom left of the Hertzsprung-Russell Diagram. They are intensely radiating stars of very small surface.The material they consist of is in fact so tightly packed that although on average their mass is the same as that of the sun, they are no larger than the earth. On earth, 1cc of this stellar material would weigh 1,000kg. White Dwarfs have reached the final stage of the dwarf stars. They are stars that, having passed the Red Giant stage and burned their hydrogen to helium, collapse; their core becomes so dense that the matter becomes degenerate. The collapse leads to ever greater densities, because the core contains high proportions of heavy elements.

White Dwarfs radiate at a surface temperature of about 10,000°C, which places them in the white star category of the early spectral classes. Because of their small surface their absolute and apparent brightness is low. They are therefore difficult to detect in the sky; most of them are too faint. Nevertheless we currently know of about 1,000 White Dwarfs.

The first White Dwarf to be discovered is Sirius B, a companion star of Sirius (the Dog Star). Friedrich Wilhelm Bessel had predicted this companion in 1844 on the basis of minor periodic motions that Sirius executed in exactly the same way as if it orbited, together with another star, a common center of gravity. Bessel was, however, unable to find this invisible companion any more than that of Procyon, the principal star of Canis minor, in which he had observed a similar undulatory motion. Here, too, the companion star that caused the gravitational effect was seemingly missing. Bessel at first consoled himself with the assumption that they must be dark, extinct stars that were no longer visible but indicated their presence by their gravitational effect on their principal stars.

In 1862 the American telescope builder Alvan G. Clark, while testing a new telescope that he pointed at Sirius, found a faint star of 8.7^m close by. It proved to be the companion of Sirius that Bessel had been looking for. But years passed before the phenomenon was explained and a new kind of star, the White Dwarf, was accepted by astronomers.

What happens to a star when it is a White Dwarf? Has it reached the end of its life? Not all White Dwarfs have a surface as hot as that of Sirius B. Some have surface temperatures of only 4,000 or 5,000°C. This permits the conclusion that every White Dwarf will one day have used up both its contraction energy and the remnants of hydrogen and helium. Furthermore, the matter inside a White Dwarf is completely degenerate because there the density is naturally very much higher than is expressed by the mean density values mentioned earlier. In the center of a White Dwarf the "electron soup" probably reaches a density of 100 tons/cc. This means that electrons, neutrons, protons, and all other elementary particles are very tightly packed. The electrons "hold the fort" and offer resistance against the further collapse of the dense core of the star. This counterpressure by the electrons is independent of the temperature; electrons can therefore offer resistance to gravity even when the body of the star is completely burnt out.

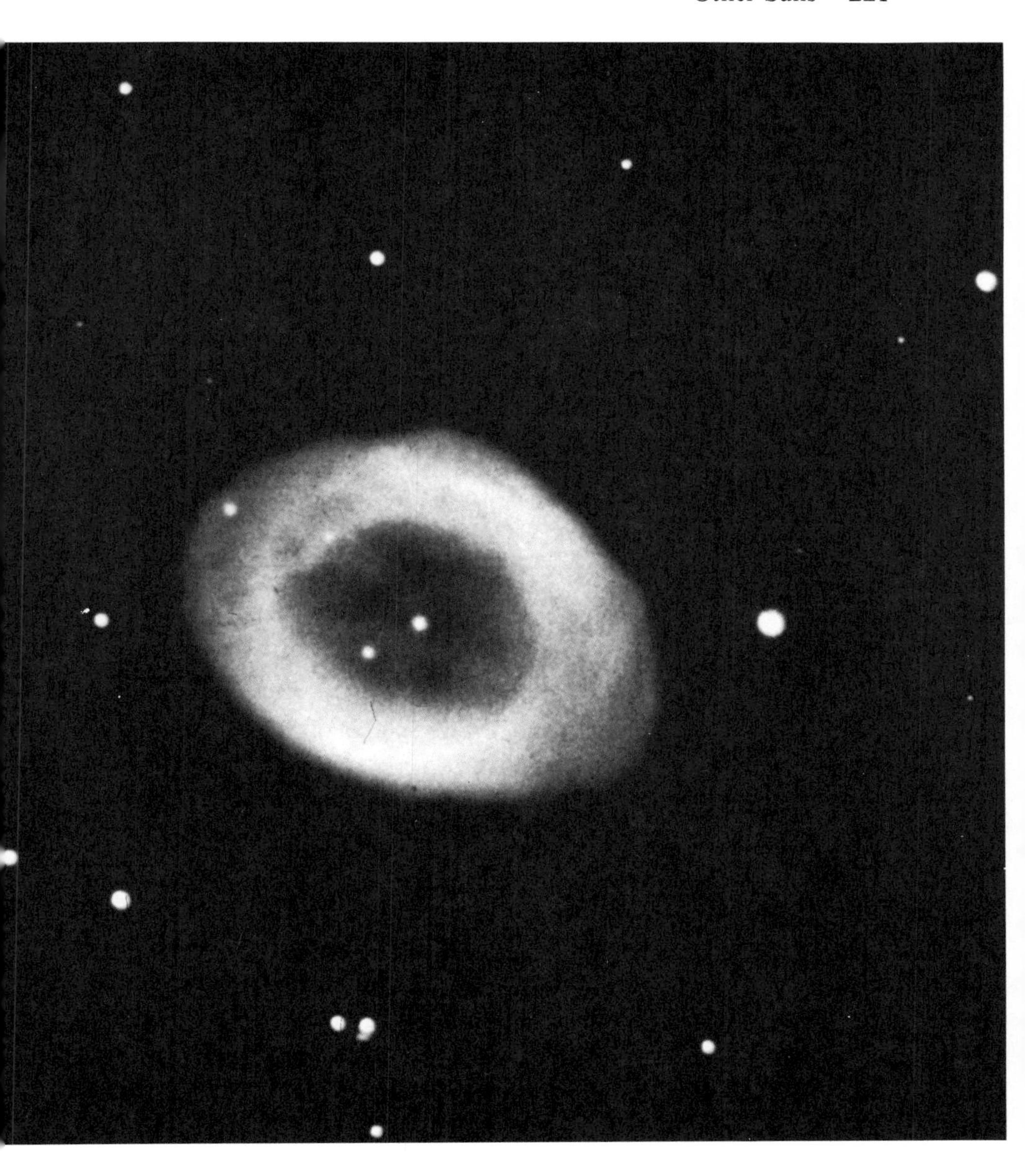

The Ring Nebula in Lyra (M57 or NGC 6720) is probably the best known planetary nebula. Its distance is 5,700 light years; the apparent magnitude of its central star is 14.7^{m}, of the object as a whole 9^{m}. The central star has a surface temperature of 75,000°C, therefore emitting mainly ultraviolet light. Its gaseous envelope expands at a rate of 38km/sec.

The White Dwarfs will therefore become completely extinct one day and cool down to absolute zero temperature. They will have become Black Dwarfs, true star cinders whose life has ended. We do not know of any Black Dwarfs today simply because the universe is not old enough for the existing White Dwarfs to have cooled to absolute zero. But they will surely exist one day.

SUPERNOVAS

Let us examine the tail end of the "stable life span" of a star, the period in which the first symptoms of a star's exhaustion appear in the burning of hydrogen to helium and of helium to heavier elements.

Consider a star considerably more massive than our sun. Its mass certainly exceeds the Chandrasekhar Mass, the 1.4 solar masses established by Subrahmanyan Chandrasekhar as the limit of mass beyond which old stars will have no chance to become White Dwarfs. This star has about 7 to 8 solar masses. During the late phase of its life it explodes very suddenly with a force beyond anything known to us.

Such a star is called a supernova. Its original luminosity increases about 1 billion times within a few weeks, and it emits as much energy as our own sun emits within the space of 10 to 100 million years. In the sky supernovas appear to be newly blazing stars.

We know of three such phenomena within our own galactic system. The original name—nova (new star)—can be understood only in historical context. Because no star had previously been noticed in the localities where these objects flared up, it was thought that they were new stars. But we must distinguish between two kinds of explosion, which have nothing in common except the tremendous eruption of light. Novas are stars whose brightness increases by seven to sixteen magnitudes; supernovas are explosions tens of thousands to a million times more violent than those of novas.

The three supernovas of historical times exploded in 1054, 1572, and 1604. Chinese and Japanese records of the time contain detailed descriptions of the first event. A star in the constellation Taurus suddenly flared up and was clearly visible even during the day. This indicates a magnitude of -5^m or -6^m. The second supernova exploded in 1572 in the constellation Cassiopeia; and at a magnitude of -4^m it was only slightly fainter than the 1054 supernova. The 1604 supernova exploded in Ophiuchus and was observed in great detail by Johannes Kepler.

In the location of the 1054 supernova a gas nebula can still be detected. It is expanding at a speed of 1,100km per second (almost 100 million kilometers per day). It is invisible without a telescope and was discovered only in 1731 by British astronomer John Bevis. The Irish astronomer Lord Rosse made drawings of and investigated this nebula with his large reflecting telescope and, because of its appearance, named it the Crab Nebula.

If we count back on the basis of the known rate of expansion of gas masses to find when these masses must have been concentrated around a single point, we arrive at 925 years ago—1054. The enormous speed at which these masses of gas expand indicates that such a supernova explosion is a different process and much more intense than the ejection of gas in a planetary nebula. Astronomers used the same methods in areas of the sky where they found cosmic masses of gas that seemed to be the remnants of a supernova. These calculations invariably made sense and therefore confirmed the suspected connection. But the advent of radio astronomy has offered another possibility for identifying likely sites of supernova explosions.

Various hypotheses exist about the origin of supernovas. Many scientists assume that supernovas are the result of explosions of old stars that have lost all reserves of energy and whose core contains mainly heavy elements (up to iron) produced by the previous burning of helium. According to one hypothesis this iron is decomposed again; what occurs is nuclear *fission* which *consumes* energy. This reaction is opposite to the one described earlier. But fission leads to the instant collapse of the star, which falls back on its reserve of contraction energy. The reaction is so explosive that the single supernova can, for a short time—a few months, perhaps—become as bright as the 100 billion stars of an entire galactic system combined. The outer portions of the star are ejected as gigantic eruptive clouds of gas. The collapse, which ends in a neutron star if not in a Black Hole, is so powerful that heavy elements—up to uranium, the heaviest element—are built up. Together with the ejected shell material they enter interplanetary space.

This theory is attractive because it also explains where the heavy elements in space come from. In the beginning there was only hydrogen and helium. Elements such as beryllium, lithium, etc., were added only gradually through nuclear transformations. The heavier elements, however, were never produced except during explosions of stars—supernovas. But this indicates that the later stars formed from masses of cosmic gas the more heavy elements they contain. The existence of our earth in its present chemical composition would be due to the supernovas, for without heavier elements life would not be feasible. At least oxygen and carbon are necessary. Can supernovas be considered the source of the "raw material" for man, animals, and plants? This, according to present-day ideas, is a quite acceptable supposition.

There are, of course, other hypotheses about the origin of supernovas. One theory, for instance, suggests that the supernova eruption is the result of the ignition of a new nuclear fire inside the star. As with the helium flash in Red Giants, a spontaneous ignition of carbon fused into heavier elements is supposed to occur in supernovas. Such a carbon flash (possible only in a star of 6 to 7 solar masses with a core of degenerate electrons, carbon, and oxygen at a nuclear density of 10^8 to 10^{10} g/cc) would generate a powerful shock wave, which would detach the outer portions of the star. Only an iron core would be left, not an object of even greater density, not a neutron star.

It was discovered in the 1950s that the Crab Nebula is a powerful source of radio waves. Then, in 1963, it was found that the Crab Nebula is also a source of x-rays. It was hoped that this discovery would confirm the suspicion that the final stage of that would be a neutron star.

Neutron Stars

Neutron stars were postulated as far back as 1934 by Walter Baade and Fritz Zwicky, both at Mount Palomar Observatory. In 1939, Robert Oppenheimer calculated a theoretical model of a neutron star.

At first neutron stars were purely hypothetical objects, the final stage of collapsed Red Giants of more than the Chandrasekhar Mass. According to calculations (which subsequently proved correct) they consisted of degenerate stellar material even much denser than that of White Dwarfs. A neutron star is an object with a diameter of only 10 to 20km and a density of 10^{13} to 10^{15} g/cc. This makes it 10 million to 1 billion times as dense as a White Dwarf. This is the density of the atomic nucleus, and only in these conditions are neutrons preserved; they do not decay into protons and electrons as they would under normal conditions.

The temperature at the core of neutron stars is about 1 billion degrees. Neutron stars also have an atmosphere, which is only a few centimeters deep and consists of hydrogen, iron, and some other metals. Its temperature is several million degrees, so that neutron stars emit mainly x-rays and only small amounts of short-wave visible light. They also emit radio waves. The radio emission of the Crab Nebula, however, proved to be an erroneous assumption: it came from the masses of gas of the remnants of the supernova itself, not from the suspected neutron star.

Another discovery, made by British radio astronomers in 1967 solved the problem. Anthony Hewish and Jocely Bell of the Mullard Radio-Astronomical Observatory in Cambridge, England, found a new kind of radio source in space. They heard signals that were repeated at intervals of 1.33730109 seconds and that lasted 1/20 of a second. At first they assumed these pulses could only be radio signals from intelligent inhabitants living on planets of other suns. But it was soon realized that these signals could not have been sent by other beings living in space because they were far too powerful. Hewish discovered other such emitters and found that the sources were point-shaped, and that they were radio stars. He called them pulsating stars—pulsars. A scan of the sky resulted in the discovery of other pulsars, mainly in densely populated regions of our galactic system. The decisive breakthrough came in 1968. A pulsar was found in the Crab Nebula at a distance of 6,300 light years; it has a period of 0.033 seconds, the shortest pulsar period so far recorded.

The American astronomer Thomas Gold, well known for his ingenious ideas, thought that pulsars could be neutron stars because only a star as dense and small as a neutron star could rotate on its axis fast enough to produce such short pulses. The powerful magnetic field of a neutron star, argued Gold, could establish a directional emission of the radio waves, like a rotary lawn sprinkler distributing water in a predetermined, constantly sweeping motion.

This theory proved plausible. We now know of several hundred pulsars, and it is fairly certain that they are all neutron stars. Pulsars, by the way, emit not only radio waves, but in flashes, similar to those of a lighthouse, visible light. It has also been found that neutron stars emit not only radio waves, short-wave ultraviolet, and visible light, but x-rays.

X-Ray Stars

X-ray astronomy is another new branch of astronomy that can be practiced only through research using high-altitude rockets and satellites. X-rays are held back by our atmosphere and are therefore impossible to observe on the surface of the earth. As a result x-ray astronomy has been established only since the beginning of the 1960s, when the search for x-ray sources in space began, first with high-altitude rockets and later with satellites. The first such point source—an x-ray star—was discovered in the constellation Scorpio in the summer of 1962. More than 100 x-ray stars were found by a special x-ray satellite launched into orbit around the earth at the end of 1972. Even now the nature of these objects is only partly understood. Some of them are in our own galaxy; others radiate from the direction of other galactic systems. The radiation intensity of some x-ray stars fluctuates considerably; others emit pulses comparable to those of the pulsars. Still others pulsate for a certain period, are completely "extinguished" for a time, then repeat their performance. Here we are dealing only with one type of x-ray star, those identical to neutron stars. This star emits pulsating x-rays, because this x-radiation is affected by the intense magnetic fields of the neutron stars. And the rapidly spinning star is also like the beacon of a lighthouse; flashes reach a point only when the beam sweeps across it.

There is a third pointer to neutron stars. They are sources of cosmic gamma radiation, that is, radiation of even shorter wavelengths than those of x-rays. Gamma-ray astronomy has opened up yet another branch of astronomy. The gamma-ray pulses are of much higher energy than the radio pulses of the neutron stars. A detailed explanation of the observations made in this field has yet to be found.

But we must mention one more phenomenon of the pulsars in this context; the pulse frequency of the neutron stars changes over time, abruptly in some, gradually in others. Is this connected with a change in the fast rotation of these objects—the only energy source left to a neutron star? If so, this would allow the conclusion that the fastest-rotating neutron stars are the youngest, the slowest-rotating ones the oldest. But what is the significance of the fact that no pulsars with rates beyond 4 seconds have so far been discovered?

The slowing down of the rotation of the pulsars is an established effect. That of the pulsar in the Crab Nebula slows down by 36.48 billionths of a second per day. Other, obviously older pulsars exhibit a similar slowing-down phenomenon. This is paralleled by a decrease in energy emitted by the pulsar. But what ultimately happens to a neutron star when its rotational energy has been used up? Will it be a dead neutron star that can no longer be detected? a star corpse?

THE MYSTERY OF BLACK HOLES

Many questions about supernovas, White Dwarfs, neutron stars, and pulsars are unsolved. But one thing is certain: they are not hypothetical objects, postulated by theory without being confirmed by practical observation. Unless our entire idea of the universe is hopelessly wide of the mark, White Dwarfs, neutron stars, supernovas, and pulsars do exist.

For Black Holes, however, there is no incontrovertible proof. Yes, theory compellingly suggests their existence, but we have only theory—no proof. And what is more, the theory claims that no such proof can ever be obtained. The Black Hole is the strangest, most bizarre, and perhaps most contradictory structure conjured up by modern astrophysics. To the layman the term itself suggests the occult rather than sober science. In fact, in England in 1973, John Taylor's discussion of Black Holes was part of a series called "Occult and Supernatural."

We have seen that when a star of less than the Chandrasekhar Mass becomes unstable at the end of its life, it collapses into a White Dwarf. If its mass is greater the collapse results in a neutron star. But there is a further limit on the scale of the neutron stars. A star heavier than 3.2 solar masses suffers such a violent collapse during its instability phase that even the densely-packed neutrons cannot withstand it. It passes through the stages of White Dwarf and neutron star incredibly quickly. This massive star becomes smaller and smaller; its density becomes greater and greater. The White (or later Black) Dwarfs, of about the same size as the earth, have a mean density of about 20kg/cc. The neutron stars (later black neutron stars) have shrunk to a diameter of 10 to 20km; their mean density is about 1.5 million tons per cc. But a massive star at the end of its life races through this phase; it drops, as it were, into a bottomless pit as far as its density is concerned, and its condensation proceeds unchecked. How dense it is at the final stage of the Black Hole is beyond imagination.

Black Holes, sometimes called collapsars, are stars whose collapse is infinite. The gravitational force on their surface increases toward the value of infinity; the volume of a Black Hole shrinks toward zero. The gravitational force of a Black Hole is so enormous that a ray of light can never leave this structure; any ray of light incident upon it, any object falling into it, is lost for all time.

How strong a force is needed to overcome the gravitational field of a body depends on the mass and density of this body. Space travel has taught us that a body that wants to leave the surface of the earth and escape the earth's gravitational field must reach a velocity of 11.2km/sec. On the moon escape velocity is lower because the mass and density of the moon are less than those of the earth. If the earth were no larger and no denser than the moon, artificial satellites would have been built and space travel achieved much earlier. On Jupiter (assuming for the sake of argument it has a solid surface) we would still try in vain to achieve space travel, because here an escape velocity of 60.5km/sec would be necessary. The escape velocity on the sun is higher still—617km/sec. On Sirius B, the oldest White Dwarf known to us, we would have to achieve as much as 1 percent of the velocity of light, 3,400km/sec; on a neutron star it is 200,000km/sec, 67 percent of the velocity of light.

But what happens when a star more than 3.2 times as massive as the sun passes through the stage of a neutron star and becomes progressively smaller and denser? The time will come when the escape velocity will reach the speed of light. The German astronomer Karl Schwarzschild developed a formula in conjunction with considerations of Einstein's Relativity Theory that allows the calculation of how small a body of a given mass must become to reach the density at which this bizarre situation will arise. The radius necessary for this, a function of the mass of the object, is called the Schwarzschild Radius (S_R). The formula is simple: $S_R = 2GM/c^2$ where G = the gravitational constant already known to us; M = the mass of the body; and c = the velocity of light. For our sun the Schwarzschild Radius would be 2.5km; for the earth, 0.9cm.

The surface of a Black Hole—that is, of the sphere whose radius is based on Schwarzschild's equation—is called the event horizon. This is, when defined by an inversion of the term, the horizon within which nothing that we could detect or prove happens.

Within the last few years probably more speculation has been indulged in about the Black Holes than about any other object of astrophysics. Are they really cosmic graves in which time and space have ceased to exist? Is the Black Hole the ultimate state toward which the entire universe is inescapably propelled? These are questions connected with the evolutionary history, structure, and future of the universe.

The Universe in Space and Time

GALACTIC SYSTEMS

On our excursion through the universe we started at our earth, investigated the sky, learned to distinguish between planets and stars, examined the moon and the individual planets, and drew conclusions about their development. Our sun was revealed as a typical star. We saw that although many stars differ from the sun and from each other in many details, they are each radiant suns. We questioned the birth and death of stars and were introduced to the hypotheses about their beginning and their end.

We also found that the many stars we can see—and many others we cannot see with the naked eye—in the night sky belong to one huge system we call the galaxy. We learned that this system consists of about 200 billion stars, is lens-shaped. It has a diameter of about 100,000 light years and a thickness of about 16,000 light years. We met many objects within this galactic system: stars, dark clouds, masses of dust and gas, matter, and fields of force. It was possible to mention only in passing objects such as binary and multiple stars, globular star clusters, and variable stars. They, too, are important parts of our universe. But this book is not concerned with details; its objective is a grand survey, an explanation of the major structure of the universe with the aid of a few examples.

Our investigation was deliberately concentrated on our own galaxy, but we know that there are many such galactic systems at unimaginable distances. As early as the seventeenth century astronomers knew of misty patches in the sky; they called them nebulae without distinguishing between different types as we do today. In this context we are concerned exclusively with a group of objects that look like nebulae and that the philosopher Immanuel Kant in 1755 suspected were not the common gas nebulae of our galactic system. Kant thought the designation "nebula" was inappropriate for this kind of object because he considered them very distant accumulations of large numbers of stars, galactic systems beyond our own galaxy. It was of course a long time before it became possible to substantiate and confirm this hypothesis.

We know now that our galactic system is only one of 200 billion such systems, each containing 100 billion suns that populate the observable universe. Because of their appearance from the earth these objects are also called spiral nebulae. This description, too, is not entirely correct. Only about 80 percent of these nebulae display a spiral shape; in the other 20 percent the stars are not arranged in a spiral. Nor is the description "extragalactic nebula" completely acceptable, because the Milky Way is, after all, just such a system of stars. Within the last few decades the term "galaxies" has become accepted for these objects.

A classical paragon of all spiral nebulae is the Andromeda Nebula, designated M31 according to a catalog of nebulae compiled by French astronomer Charles Messier in 1771. For a long time it was doubted whether this structure was indeed another galaxy. But in 1885 the flaring up of a supernova in the Andromeda Nebula, whose apparent brightness indicated that the Andromeda Nebula is far beyond our own galaxy, supplied the first evidence. Further proof was provided by photographs taken toward the end of the nineteenth century. These indicated the spiral structure of the nebula.

But a definite confirmation of the extragalactic nature of the Andromeda Nebula was only established in 1925 when Edwin Hubble succeeded in resolving parts of the Andromeda Nebula into individual stars. By comparing the apparent brightness of these stars with those of stars in our own galaxy he arrived, initially, at very rough estimates of the distance of the Andromeda Nebula. He used for this purpose mainly a type of star known as a Cepheid.

Cepheids are very luminous supergiants whose brightness fluctuates periodically. In 1912 the American astronomer Henrietta Leavitt investigated the Cepheids in the small Magellanic Cloud. The Large and the Small Magellanic Cloud are two galaxies near our own, in front of it as it were. They are so close to the South Pole of the sky that one has to travel almost as far as the equator to be able to observe them. Leavitt discovered during her investigations that the period of brightness variation of the Cepheids in the Small Magellanic Cloud is the longer the brighter the star. Since the Small Magellanic Cloud is 165,000 light years away we can consider distances between the stars that form it negligibly small. We can also regard these stars as equidistant from us—this makes their apparent brightness a measure of absolute brightness or luminosity. It must be possible, then, to conclude from the period of light variation of a Cepheid its distance once the distance of one Cepheid is known.

Hubble found Cepheids in the Andromeda Nebula and in other galaxies in our neighborhood. He thus determined the distance of the Andromeda Nebula as an astonishing 1 million light years, and those of other spiral nebulae as up to 7 million light years. It was known that our galaxy had a diameter of 100,000 light years, so Hubble's figures were proof of the extragalactic nature of the spiral nebulae.

Hubble also conducted investigations into the number of galaxies in the universe. He began to collect statistics of the spiral nebulae by photographing a large area of the sky through it with the 2.5m telescope at Mount Wilson. Each of his exposures lasted one hour. He counted about 44,00 spiral nebulae on his plates, a few hundred per square degree of the sky. Their distribution seems to be uniform everywhere except in the direction of the Milky Way. In this direction the numerous dust and dark clouds that form part of our galactic system obscure the view. Allowing for a correction factor derived from this observation Hubble calculated that the telescope would record about 3 million galaxies throughout the sky on photographs exposed for one hour.

Hubble examined even the faintest, most distant galaxies. By comparing their brightnesses with those of the spiral nebulae whose distance he had been able to determine he found that the 2.5m telescope reaches about 600 million light years out into the universe.

Spiral Nebula M51 is in the constellation Canes Venatici (Hunting Dogs) near the end of the tail of Ursa Major. Because one of its spiral arms extends upward to the smaller appendix, it was assumed that they were neighbors in space. Computer data now suggests that M51 is a minor galaxy drifting past and attracting matter from the major partner through its gravitational force.

THE EXPANDING UNIVERSE

In 1913 Vesto Melphin Slipher made a strange discovery. He had investigated the spectra of a few spiral nebulae and found that their spectrum lines were often displaced from their normal position toward red, the long-wave end of the spectrum.

The spectra of spiral nebulae are largely the superimposed spectra of the individual stars constituting these islands in space. We can therefore learn from the spectrum of a spiral nebula, for instance, what stars in this system dominate in brightness and number. Especially in the arms of the spirals we find many blue supergiants, so an early spectrum is recorded there. We also distinguish between various types of spiral nebulae according to their spectra, as we classify according to appearance.

But what aroused Slipher's interest was a shift in the lines of spiral nebulae spectra. Line shifts in the spectra of stars were by no means unknown at the time; they were interpreted in terms of the Doppler effect—a change in frequency of both light and sound waves discovered by Austrian physicist Christian Johann Doppler. The change in wavelength occurs when the object producing it approaches or recedes from an observer. This can be most clearly demonstrated with an automobile sounding its horn. As it approaches us we hear a high pitch; the moment it reaches us and begins to move away the pitch changes to a lower value. The sound waves are compressed as the automobile moves toward us, the wavelength becomes shorter. In other words, the frequency or number of wave crests per unit of time, increases. When the automobile moves away from us, the wave length increases, the pitch becomes lower, the frequency decreases.

The process also occurs with electromagnetic waves, which include the region of visible light. Doppler expressed this in the equation:

$$v = \frac{\triangle\lambda}{\lambda \times c}$$

In this equation v = motion of the object toward or away from us, the radial velocity; λ = wavelength; $\triangle\lambda$ = wavelength change; and c = velocity of light. If the equation is to be used for the calculation of the radial velocity of an acoustic source, c would be substituted by the velocity of sound.

The examination of the spectra of stars and the shift of spectrum lines thus provides information about the radial motion of the stars. If a shift occurs toward the violet short-wave part of the spectrum, the object is approaching us. A shift toward red is a sign that the object is moving away from us. Doppler's formula even provides quantitative information about the extent of the radial motion.

Stars exhibit, as expected, the most varied shifts of their spectrum. Some display no Doppler shift at all (which means that these stars have no radial motion relative to us); others show violet. But Slipher's spiral nebulae exhibit mainly red shifts. Interpreted according to the Doppler effect, they are moving away from us. Slip-

This spiral nebula, NGC 4565, is in the constellation Coma Berenices (Berenice's Hair) and is seen here edge-on. It has a well-defined center and prominent dark areas—dust clouds—which are also found in our galactic system.

her revealed this discovery at a meeting of the American Astronomical Scoiety in 1914. Nobody was yet in a position to assess the significance of what he had to say, but the audience—Edwin Hubble among it—was aware of the importance of Slipher's findings.

Slipher continues his investigations about the Doppler shifts in the spectra of spiral nebulae. By 1925 he had determined the radial velocities of forty-two galaxies. Almost all were moving away from the earth at great speed. Hubble and his colleague Milton Humason continued Slipher's work.

In 1917 Albert Einstein published his General Theory of Relativity. The Dutch astronomer Willem de Sitter found a solution of the formulas of this general theory, according to which the universe is expanding continuously. Einstein, owing to a simple miscalcula-

Example of a very open spiral nebula, seen directly from above.

tion, had overlooked this consequence of his Relativity Theory; but a Russian mathematician, Alexander Freidmann, arrived at the same conclusion of an expanding universe. Hubble was aware of this theoretical idea of an expanding universe, and he appreciated the connection. Indeed, it was de Sitter's hypothesis that made Hubble continue Slipher's work. If, he argued, the distances and the red shifts of a large number of spiral nebulae were determined, the theory of the expanding universe could be either proved or disproved.

Together with Humason he began to measure spiral nebulae. He determined the distance of these objects by all available methods; Humason measured the red shifts in their spectrum. Their tool was the powerful telescope on Mount Wilson. Hubble's observation took him to distances of 100 million light years. This made him the first astronomer to realize the enormous size of the universe. He then turned to Humason's results of the red shift measurements. Practically all the spiral nebulae displayed; the great majority apparently were moving away from our corner of the universe.

Hubble discovered that the velocity at which the distant spiral nebulae recede from us is greater the more remote these objects are from us. This law of the expanding universe, postulated in theory by de Sitter, was empirically proved by Hubble's and Humason's observation.

Einstein found the idea of an expanding universe sinister because it implied that it must have had a beginning and that there must have been a point in time when expansion began. Only in 1930, after seeing Hubble's spectrograms, did Einstein become convinced that we live in a dynamic, not a static, universe. In this expanding universe every point moves away from every other point. The center of this universe is nowhere and everywhere. No star system is privileged over another.

In 1948 a new telescope was installed on Mount Palomar. This 6m reflector was the world's largest telescope for several years as was Mount's Palomar's 5m reflector before it. Hubble repeated the sky-scan for spiral nebulae with the new instrument. Again he exposed his plates for one hour and reached islands in space twice as far away from us as the faintest objects he was able to reach earlier—1,200 million light years. Even at these enormous, unfathomable distances, the frequency of distribution of the spiral nebulae was as high as at lesser distances. Did this mean that the universe is infinite? Are there an infinite number of galaxies?

Here an interesting counterargument arose. If the universe contained an infinite number of star systems that emitted an infinite quantity of light, how could the sky be dark at night? Should one not assume that the universe and the number of galaxies is finite after all? The law of the expanding universe refutes this argument. It can be shown that the light emitted by the galaxies is attenuated—diluted, as it were—by the movement of each galaxy away from the others, because the quantity of light emitted by every galaxy has to fill an ever-increasing area in the universe. As a result the quantity of light is more widely dispersed than it would be were

Two examples of red shift in the spectra of the spiral nebulae.

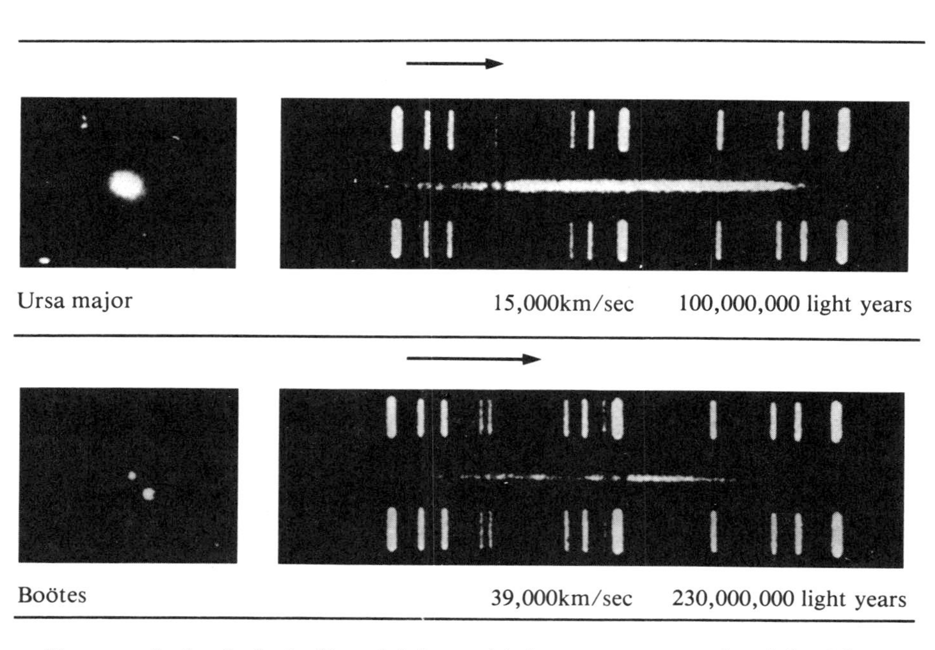

Top: a spiral nebula in Ursa Major, with its spectrum on the right. The arrow indicates the extent of the line shift: a radial velocity of 15,000km/sec, according to which the distance is 100 million light years. *Bottom:* a galaxy with its spectrum. The distance is 230 million light years; velocity of recession, 39,000km/sec.

the galaxies at rest. This explains why the night sky is not radiantly bright with innumerable galaxies radiating an infinite quantity of light. During its short history Hubble's constant—the numerical relation between the distance of the spiral nebulae and the measure of the red shift—has been emended three times. The first correction became necessary after 1930, when it was realized that interstellar absorption—the attenuation of light by masses of dust between us and the spiral nebulae—had to be allowed for. The second correction, in 1952, was due to Walter Baade's discovery that there are two types of Cepheid; and the third, in 1960–70, was called for by several other factors. The upshot of all this is that the distance of the Andromeda Nebula is not, as originally assumed, just under 1 million, but 2.3 million light years from earth.

Today the value of Hubble's constant is 55km/sec velocity of recession per megaparsec distance, or about 17km/sec per million light years. Generally, $v = H \times r$. In this equation, v = mean velocity of recession of the galaxies; H = Hubble's constant (55km/sec/megaparsec); and r = distance (also in megaparsec). The velocity of recession of the spiral nebulae is so high that relativistic effects must be allowed for, statements of the relativity theory about velocities in the regions of that of light, where conventional considerations are no longer valid. The following equation describes the Hubble effect and the line shifts in the spectrum:

$$\frac{\Delta\lambda}{\lambda} = \frac{\sqrt{1 + v/c}}{\sqrt{1 - v/c}} - 1$$

Here, v = velocity of recession, and c = the velocity of light.

The continuous expansion of the universe also implies that billions of years ago the galaxies must have been much closer together. We can indeed count back, if we accept the expanding universe, to the time when all galaxies originated at a single point. According to calculation this must have been 15 to 20 billion years ago. Is 15 to 20 billion years the age of the universe? Was this when it all began? And if so, what existed before? If we subscribe to the hypothesis of a universe that began at age zero, we cannot ask these questions; in terms of physics it is pointless. A statement about it is no more possible than is a meaningful answer to the pointless question of how bodies behave at a negative absolute temperature.

Our theory of the expanding universe suggests that there was a zero point when the entire matter of the universe was concentrated

This picture contains many far-distant spiral nebulae—a "nest" of nebulae. It was taken with the 5m telescope at Mount Palomar and shows three to four times as many nebulae as would a photograph of the same area in identical conditions taken with the 2.5m telescope at Mount Wilson. Even the brightest star in this picture (top right), much over-exposed, would have to be sixteen times brighter to be visible.

at one point in space. This primeval condition of the universe is beyond imagination. But a very detailed theory exists about what happened after zero point. This theory of the origin of the universe is called the Big Bang hypothesis and postulates a universe that started with a tremendous explosion.

BIG BANG OR STEADY-STATE?

Steven Weinberg, in *The First Three Minutes,* has impressively described this story of evolution. It begins with a universe that consisted of a dense soup of matter and radiation at a temperature of 100 billion degrees Kelvin and that expanded, exploded, at tremendous speed. To describe the further development in detail is beyond the scope of this book. Suffice it to say that it was an explosion in the sense that all parts moved away from each other and from a common point of origin at high velocity and that the initial temperature dropped dramatically. After 34 minutes and 40 seconds this universe had "cooled" to 300,000°C. Then nothing remarkable happened for 700,000 years; expansion and cooling continued; atoms were produced, 22 to 28 percent were helium, the rest hydrogen. More and more stars evolved from this material. Weinberg's figures are the result of solid, complex considerations of nuclear physics and astrophysics; every one of them can be substantiated and verified.

The Big Bang hypothesis suggests that the quantity of matter in the universe, the number of galaxies, is finite. The universe came into being with a finite quantity of matter, emerged from a mighty gas cloud of hydrogen and helium, and dispersed into individual clouds which formed galaxies. From these galactic gas clouds developed the stars as we know them today. There must also be an end to the universe; it cannot be infinitely large. The universe is not infinitely large, yet is has no limits. It is, to quote Einstein, a four-dimensionally curved space.

Its significance can be interpreted only in terms of mathematics. Man, like his entire environment, is an object of three-dimensional space. The fourth dimension is beyond our grasp. We can only deal with it mathematically and deduce from the results of these investigations the validity or otherwise of this abstract concept of four-dimensional space. But all the criteria established so far confirm that Einstein's theory of four-dimensional space was right. We can formulate analogous ideas about this space. It is possible, for instance, to compare our relation to four-dimensional space by comparing a two-dimensional creature and a three-dimensional world. This will lead a rational being to some startling discoveries which it will not be able to grasp, but with which it can operate mathematically, and deduce from certain observations in its world that the mathematical description is correct. For example: a two-dimensional creature on a globe would, for instance, determine that although its world is finite it has no limits. For wherever our creature moves in this world it will never reach a limit; it will merely return to its starting point. Its world is curved into a third dimension.

This picture of the sky in x-ray light was taken by the fully automatic NASA satellite HEAO-2, also called the Einstein Observatory. The photograph shows a newly discovered quasar (quasi-stellar radio source) at a distance of 10 billion light years. The most distant quasar found so far is 15.5 billion light years away, at the edge of the known universe.

Another experience this creature would have is that the sum of the angles in a triangle may be larger than 180°—a fact incomprehensible to it, yet verifiable. It is clear to three-dimensional creatures because we know the spherical triangle—a triangle on a sphere—in which the sum of the angles is naturally larger than 180°.

We can make similar observations. They all show that space is four-dimensionally curved, and that the curvature is related to the quantity of matter present in various regions of space. But there is a second theory about the universe, the Steady-State hypothesis. This view holds that the universe is eternal in time and space and as a whole is unchangeable. It has no beginning and no end; it is continuous.

According to the Steady-State hypothesis the universe was not born 20 billion years ago but has always existed and will always exist. To the extent that it expands, new matter is being formed so the density of the universe remains steady. Calculation shows this to be a theoretical possibility. All that would be necessary to substantiate it would be the creation of a new hydrogen atom per 4 cu.km per year. From those hydrogen atoms new galaxies would form continually, new stars would be born, old ones would die. This universe would look exactly today as 20 or 100 billion years ago, and will look exactly the same in another 20 or 100 billion years.

Astronomers have tried for a long time to find a criterion that will enable them to decide which of the two possibilities is correct. According to one consideration, in a Big Bang universe the most distant galaxies should look younger than galaxies closer to us. But nobody knows the difference between a young and an old galaxy. If all the far distant galaxies were young this would prove the Big Bang theory correct. But if some were old, others young, this would point to a continuous universe.

The solution was provided within the last few years by a completely unexpected branch of astronomy. Radio astronomy provided concrete proof that the universe began as the explosion of a cosmic "hydrogen bomb." In 1965 two scientists, Arno Penzias and Robert Wilson, discovered a radio emission that appears to come from all directions of the universe. No single object could be identified as its source. It turned out that the two scientists had discovered the basic radiation that astrophysicists had postulated as early as 1948 if the Big Bang theory was correct. For this theory has long suggested that the universe must have been a huge ball of fire within the first seconds after the primeval explosion. To the extent that the universe expanded and cooled the brilliance of this fire ball diminished, but the radiation has never completely disappeared. It became obvious that it was this diffuse radiation from the original state of the universe that Penzias and Wilson discovered. And although the theoreticians who preferred the Steady-State model tried desperately to establish another reason for the presence of the primary radiation, they failed. The radiation detected by Penzias and Wilson has the exact frequencies and intensity that must have been produced during the great explosion of the universe. As a result as we can now assume with almost absolute assurance that the universe had a beginning in space and time. But what about its future?

If we follow the Big Bang theory we must expect that the universe will reach a final stage at some time. Creation and dissolution, life and death, would therefore be the law ruling not only man, animals, plants, and planets, but all stars and, ultimately, the entire cosmos. Does it mean that even the awe-inspiring universe is merely an episode, before and after which there is eternal timelessness? Questions such as these transcend our powers of imagination.

Yet another hypothesis, although so far unsupported by observation, is worth consideration. It is the idea that the radius of curvature of our universe will not increase to all eternity—the distribution density of galaxies will not continually decrease—but the universe will reach a limiting size and, after a period of steadiness, begin to contract until it returns to the state of the Big Bang—matter so tightly packed that it will form an almighty ball of fire as it did originally (if the Big Bang theory is correct). There will be a repetition, then, of an event that happened 20 billion years ago: a new creation of the world, a new formation of chemical elements, of stars, of galaxies.

Such a universe would be conceivable if the distribution density of the spiral nebulae exceeded a certain calculable value, when gravity could lead to an end of expansion and to a reversal of the entire universe.

Many astronomers therefore concern themselves with the question of the number of galaxies in the universe and their density of distribution. At the moment all quantitative results point to a steadily expanding universe—matter in the universe is simply not dense enough to reverse the expansionary motion of the universe. This, at any rate, is the verdict of the data we have at the moment; but they are provisional values. Perhaps someone will, in the coming years or decades, discover correction values that have to be applied. Perhaps the concept of a pulsating universe, oscillating between two limiting states, is the right one after all. Perhaps the universe will, after all, last forever.

References

Jastrow, Robert. God and the Astronomers. 1978.

Plutarch. The Face in the Moon's Disk.

Freisleben. Galileo Galilei. Stuttgart, 1956.

Ley, Willy. Die Himmelskunde (Astronomy). Duesseldorf, 1965.

Plassmann. Himmelskunds (Astronomy). 1913.

Littrow and Stumpff. Wunder des Himmels (Magic of the Sky). 11 ed. 1969.

Picture Credits

pp. 21, 172, drawn by Roland Zahn, from the German Publishing Institute; *p. 8,* Arthur Grimm, Berlin; *p. 4,* E. Krug, Berlin; *p. 181 (bottom),* NASA, Washington; *pp. 179, 180, 181 (top),* Sacramento Peak Observatory, Geophysics Research Directorate, ABCRC; all other illustrations, Werner Buedeler Pictorial Archive, Thalam, Germany.

Index